PLANT MYSTERIES
A *Scientific Inquiry*

If you talk gently and in a pleasant tone of voice to plants, will they grow better? If you play soft, classical music near the plants instead of hard rock, will it make any difference to the plants? Do they show fear and respond to pain? Do plants have "feelings" the way you and your friends do? Many experiments have been conducted to try to "prove" that plants do feel things. But have these tests been accurate and scientific? While they acknowledge the more sensational theories about the sensitivity and responses of plants, the authors believe there may be scientific, biological reasons why plants act the way they do. Plants do exhibit strange, mysterious behavior—is this a sign of "humanlike" traits or can it be explained logically and scientifically? Read on and form your own theories about the secret life of plants.

BOOKS BY MICHAEL J. CUSACK
PLANT MYSTERIES: A Scientific Inquiry
IS THERE A BERMUDA TRIANGLE?

PLANT MYSTERIES
A Scientific Inquiry

ANNE E. CUSACK
and
MICHAEL J. CUSACK

photographs and drawings

JULIAN MESSNER
NEW YORK

STAR OF THE SEA GRADE LIBRARY

Copyright © 1978 In Trust for the benefit of
Deborah Anne Cusack, Deirdre Ruth Cusack, and
Jennifer Ellen Cusack.
All rights reserved including the right of
reproduction in whole or in part in any form.
Published by Julian Messner, a Simon & Schuster
Division of Gulf & Western Corporation, Simon &
Schuster Building, 1230 Avenue of the Americas,
New York, N.Y. 10020.

Manufactured in the United States of America

Design by Irving Perkins

LIBRARY OF CONGRESS CATALOGING IN PUBLICATION DATA

Cusack, Anne E
  Plant mysteries.

  Bibliography: p. 149
  Includes index.
  SUMMARY: Describes human-like sensitivity and responses of plants, such as perceiving and reacting to people's thoughts and feelings, and presents scientific reasons for such behavior.
  1. Plants—Irritability and movements—Juvenile literature. 2. Parapsychology and plants—Juvenile literature. [1. Plants—Irritability and movements. 2. Parapsychology and plants] I. Cusack, Michael J., joint author. II. Title.

QK771.C87   581.1   78-12665
ISBN 0-671-32897-2

Contents

1. The Surprising Kingdom — 9
2. People Talk to Plants — 21
3. But, Do Plants "Talk" Back? — 31
4. Encounter with a Lie Detector — 37
5. ... and Related Experiments — 47
6. Leading to a "Secret Life" — 55
7. Do Plants Have Feelings? — 65
8. The Experts Take a Turn — 75
9. A Certain Unity of Life — 83
10. Responses and Adaptations — 91
11. Light ... and Gravity — 101
12. A Chemical Awareness — 113
13. Plant Hormones — 121
14. Plants and Sound — 129
15. A Question of Sensitivity — 141
    Suggested Further Readings — 149
    Index — 151

PLANT MYSTERIES
A Scientific Inquiry

CHAPTER 1
The Surprising Kingdom

YOU STEP AMONG some delicate-looking plants on a summer day. Suddenly, as if by magic, the plants seem to disappear. The ground is left nearly bare.

You pause, silent and motionless. A few minutes later, the plants are "back." The field is lush and green again.

Impossible? No . . . not if the plants happen to be of the species *Mimosa pudica*, commonly known as the sensitive plant.

Mimosa pudica is a plant native to Central and South America. It has broad leaves. And it behaves in a very surprising way. At the touch of a finger, at the footstep of a small animal, or at the brush of a bee's wing, this plant seems almost to vanish.

In response to touch or sound or shock, the broad leaves of *Mimosa pudica* will furl like the folding of a fan. Then each furled leaf will drop backward and downward at the joint. What was shortly before a spreading, broad-leafed plant will appear to be little more than a

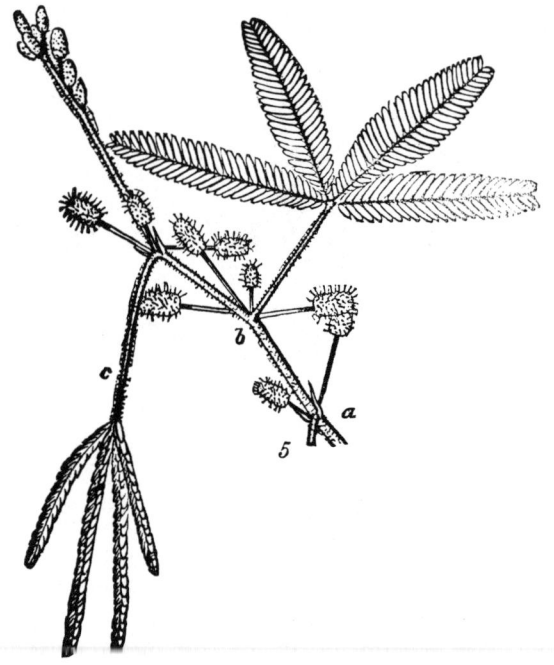

Mimosa pudica. Plants of this species are particularly sensitive to such stimuli as sound and touch. In response to noise or touch, the compound leaves of the plant curl up and appear to wilt. This is caused by a sudden change in the plant's water balance. Each leaf is actually an array of tiny leaflets on either side of a twig. There is a water-filled sensitive organ at the point where each leaflet is connected to the twig. In response to touch or sound, the water will rapidly flow out of each of these sensitive organs. As a result, the leaflets will collapse and lie against the twig.

The Surprising Kingdom

few twigs. Ground that was covered with these plants will appear almost bare.

If the *Mimosa pudica* plants are left undisturbed for a time, the broad leaves will slowly, cautiously, unfold. But any further touch or sound can cause the leaves to "hide" again.

Why do the leaves of *Mimosa pudica* fold when the plant is disturbed? Is the sensitive plant "afraid" of insects and people? Does it, or any plant, actually have feelings?

The nineteenth-century English poet Percy Bysshe Shelley may have thought so. His poem "The Sensitive Plant" describes a lovely garden wherein each flower "shared joy in the light of the gentle sun." But one plant in that garden, the sensitive plant, "felt love from leaf to root."

This lovely garden, Shelley wrote, was tended by a lady whose every "step seemed to pity the grass." She watered the plants from a nearby stream. She tied the taller plants to rods and stakes. She even picked insects and gnawing worms off the plants with her gentle fingers. Then, unwilling to kill the tiny animals, she took them to the woods and released them.

"I doubt not," said Shelley, that the flowers "felt the spirit that came from her glowing fingers through all their frame."

When the lady died, the whole garden mourned her. "The Sensitive Plant, like one forbid, wept." Its leaves

PLANT MYSTERIES

and branches drooped. Under its roots, moles and mice died.

When spring finally returned to the garden, docks and darnels—common weeds—came up once more, but the sensitive plant was a leafless wreck. The poet implied that grief was the cause of the plant's death. The sensitive plant, he wrote, "felt the sound of the funeral chant and the steps of the bearers."

The idea that plants can *know* how people feel about them is not new. Various primitive peoples believed that spirits lived in trees and certain other plants. Those people took care not to offend the spirits of the plant world by thought, word, or deed.

Members of several primitive societies today still hold such beliefs. But they are not alone in having a great sense of admiration and awe for members of the plant kingdom. Poets and saints have expressed a feeling of oneness with plant life. Even practical scientists frequently talk about their feelings of respect and wonder at the mysteries of the mighty plant kingdom.

Charles Darwin (1809–1882), the great biologist who advanced the theory of evolution, wondered whether plants might have some form of awareness. Darwin could never find any evidence that plants have nerves or brains as animals do, but he once compared the growing tip of a plant embryo to an animal's brain. Furthermore, Darwin never ceased to wonder at the complex ways in which plants adjust to their environment.

The Surprising Kingdom

Many people today consider that plants show what might be called a "certain awareness" of their immediate environment. Some plants, after all, sometimes do surprising things.

Plants open and close their flowers and leaves, often at a predictable time of day. Plants seem to "know" when spring or fall is near, and they prepare themselves for the coming season.

Many plants can be fooled by a smart farmer or florist into blooming out of season. Some plants can be tricked

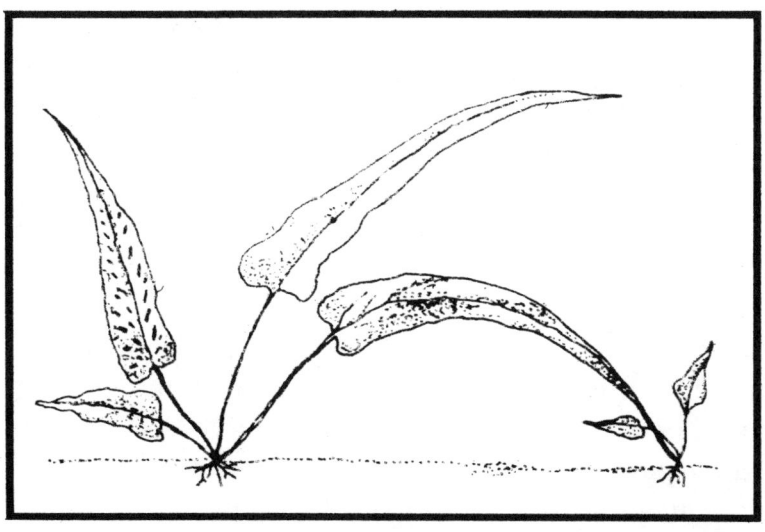

Camptosorous rhizophyllus. Ferns of this species sprout roots from the tips of their long frond or leaves. When the tip of a frond of one of these walking ferns dips into the ground, it takes root. A tiny plant grows up from that root.

PLANT MYSTERIES

into budding too early by an unseasonal thaw. But, by and large, plants seem to manage their lives with remarkable precision. Some plants appear to respond to certain conditions in the world around them.

Farmers know that if a stake is placed near a climbing bean plant, the plant will locate the stake and then start twisting itself around the stake.

Tropical orchids often resemble in shape and color the particular birds or insects the plants rely on for pollination and seed transport. Several other plant types also seem to mimic certain animals.

Some plants even seem to "walk" around. Ferns of the species *Camptosorus rhizophyllus* sprout roots from the tips of their long leaves. When a frond (leaf) tip of this "walking fern" dips into the ground, it takes root, and a tiny plant grows up from that root.

Some members of the plant kingdom appear to be positively ferocious. They eat meat. In fact, they trap their own dinners.

Best known among these carnivorous plants are the Venus's-flytraps. Such plants, which are found growing wild only in swampy areas of North and South Carolina, are equipped with snap-trap leaves perfectly suited for catching insects and other small animals.

Venus's-flytrap leaves are fringed with bristles. Inside each leaf there are a number of trigger hairs. Should an insect alight on an open leaf and touch one or more of

When this illustration first appeared in 1925, stories of man-eating trees were not uncommon. But the stories weren't true. There are no such trees. All carnivorous plants are small. The largest animal one of them could eat would most likely be a tiny reptile or amphibian. (*Garden Journal*)

these hairs, the leaf will snap shut. The bristles will interlock so that the insect cannot escape. Then juices released by the plant will smother, soften, and digest the unfortunate animal.

Botanists are aware of about 450 species of carnivorous

PLANT MYSTERIES

plants in the world. Different groups of these plant species use different methods to trap their prey. Bladderworts, for instance, use suction traps to catch extremely small animals.

The branches of a bladderwort plant have sacs like tiny deflated balloons. Each sac is sealed by a thick gluey plug. When the plant's trigger hairs are touched, a valve opens. Water flows in as the walls of the sac expand. One-celled water animals called protozoa and other tiny creatures such as insect larvae are sucked into the sac along with the water. These tiny animals are digested and provide nutrients for the plant. Plants of some bladderwort species can reset their traps by pumping the water out of the sacs and then resealing the valves.

We've mentioned just a few of the many surprising things plants sometimes do. Considering the remarkable responses of certain plants, it is not strange that many people suspect that plants are to some extent aware of what's going on around them. Nor is it surprising that, to many people, talking to a plant seems natural.

A garden flower, and especially a houseplant, is, after all, a kind of pet. Its owner may grow it from seed, root it from a cutting, or buy it from a store. In any case, the owner is involved in the plant's early growth. He usually sees it each day. He waters it and, if necessary, he feeds it.

The plant owner may tie the plant to a stick or to a small trellis as it grows taller. On a bright morning, the

The Surprising Kingdom

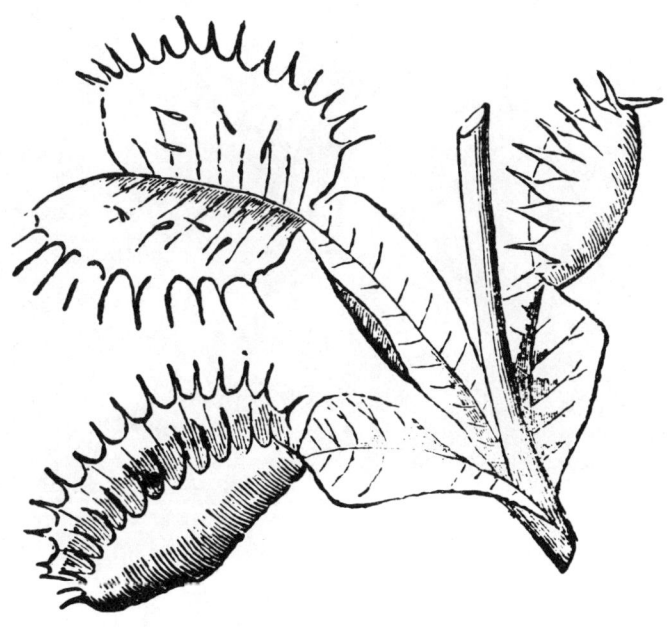

Dionaea muscipula. This species is better known as Venus's flytrap. They are perhaps the best known of the carnivorous plants. They are equipped with snap-trap leaves for catching insects and other small animals.

person may move the plant to a sunlit area. On very hot days, he may move it into the shade. Each and every day, the plant owner may spend much time caring for that plant.

If the plant grows well and blooms, the person is likely

Sarracenia purpurea. This is one of several species of carnivorous plants generally called pitcher plants. These plants employ a "pool of death" at the bottom of pitcher-shaped leaves to trap small insects. The pool contains a sweet-smelling liquid to attract insects and digestive juices with which to "eat" them. Insects attracted by the odor tend to slip down the sides of the "pitcher" into the pool.

The Surprising Kingdom

to feel proud and happy. But if the plant wilts or drops its buds, the person may worry about it. Is the plant getting too much water? Or not enough? Is it bothered by spider mites, aphids, or some other pests?

The worried plant owner may actually ask the plant what is wrong with it. Or he may try consoling and comforting the plant.

"Maybe this will help," a plant owner may mutter as he snips off a dead leaf. Soon—almost without realizing it—the person may be talking to his or her plants.

CHAPTER 2
People Talk to Plants

MANY PEOPLE TALK to their plants. Many people seem to have a special way with plants. Often the two seem related.

People with a special way with plants seem to know just when to water them, how much water to give them, and just what to feed them. The gardens of these "greenthumb people" tend to outproduce all other gardens in the neighborhood. The houseplants of these "greenthumb people" flourish and win prizes.

If you ask such people how they do it, they will usually talk about keeping the air around the plants moist and not letting the soil dry out. They may also talk about using the "right" soil and the "right" fertilizer.

"Greenthumb people" usually know a lot about plant needs and practical plant care. But is that all there is to it?

PLANT MYSTERIES

Listen in on some "plant people" as they talk about their plants.

"This one doesn't like wet feet," a "plant person" explains. "But that fellow doesn't mind at all. He's a thirsty one."

Another person says, "I give my plants a bath once a week. But what they really love is a cool shower in the summer. I take them out to the front porch and let the rain wash them off. You can almost see them smile."

"Plants get lonely all by themselves," says a third plant enthusiast. "They like each other's company. They grow better with other plants near them."

"Plants are a bit like children," says another "plant person." "They need lots of attention. You have to know their habits. You know, I sometimes catch myself talking to them."

The thought that plants need special attention is not just a notion of pet-starved, modern city dwellers. Many old-time gardeners shared this belief and talked to their plants.

In his book, *Talk to Your Plants*, gardening expert Jerry Baker describes how his Grandma Putt cared for her plants. Grandma Putt was part Indian and lived in Iowa. She believed that people were meant to live in harmony with nature. Like many other experienced gardeners, Grandma Putt combined common sense and traditional plant-care methods with careful observation

People Talk to Plants

of plant behavior. As a result, her flowers and vegetables were the pride of the county.

Grandma Putt recommended talking to plants. She spoke to them herself, says Baker. She believed that to be successful with plants, a person needs to know how to communicate with them. Along with many people of her time, she saw it as a perfectly natural thing to do.

Some very famous "plant people" talked to their plants. The list includes the famous agricultural scientist, George Washington Carver (1864–1943).

As you may know, Dr. Carver made peanuts an important cash crop for farmers of the southeastern United States. Carver realized that Southern farmers were wearing out their land by planting cotton on it year in and year out. This one-crop farming was robbing the soil of nitrogen, an essential plant nutrient. To put nitrogen back in the soil, Dr. Carver decided that farmers should plant legumes instead of cotton. Legumes—peas, beans, peanuts, alfalfa, and clover—have the ability to "fix" nitrogen from the air and return it to the soil in a form usable to other crops. The legume Dr. Carver had in mind for this purpose was the peanut, which up to that time had been used mainly for hog food.

George Washington Carver developed new uses for peanuts. As a result, many farmers were encouraged to plant peanuts instead of cotton. Today, thanks to the efforts of the "Wizard of Tuskegee," peanuts are an

PLANT MYSTERIES

important cash crop in the South. Much of what was once worn-out land in the southeastern states is healthy again.

"Fathering the peanut-butter sandwich" isn't Carver's only claim to fame. He has been called a genius in devising plant-care methods. As a child, Carver took care of his neighbors' sick plants. He would sing to the plants as he potted them in tin cans. He would often talk softly to them as he carried them into the sun.

Dr. Carver never lost this ability to commune with plants. Even as a professor at Tuskegee Institute in Alabama, he said that his discoveries came to him as flashes of insight while he walked in the woods. He often turned to the plants themselves for inspiration. Once, while staring at a peanut plant, he shouted, "Why did the Lord make you?"

Another famous plant wizard, Luther Burbank, also often turned to the plants themselves for inspiration. On his seed farm in California, Burbank developed hundreds of new varieties of trees and other plants. Many of the delicious varieties of fruits and vegetables we enjoy today are the results of Burbank's labors.

Luther Burbank was a devoted reader of the books of Charles Darwin. As he crossed or interbred plant varieties to select the healthiest and strongest for propagation, Burbank felt that he was putting Darwin's theory of evolution into practice.

Burbank also believed in the value of talking to plants.

People Talk to Plants

However, he was not always willing to admit this. He was afraid that he would be laughed at. In fact, Burbank once told a young reporter that stories about him talking to plants were ridiculous because plants haven't any brains.

Yet, with his friends Burbank shared his feeling that plants and humans could somehow understand each other. Manly P. Hall, a minister who later founded the Philosophical Research Society, said that Burbank would ask his plants to help him when he wanted them to grow in a new way. Though he was not sure they understood his words, Burbank felt that the plants could somehow grasp his meaning and respond.

Until recently, most people who talk to plants have been shy about admitting it. Like Burbank, they were afraid that people would laugh at them. Yet, like Burbank, they felt that the plants could somehow grasp their meaning and respond.

Most botanists (plant scientists), however, tended to scoff at the idea of communicating with plants. Plants have no ears. And botanists have not been able to find any type of brain or nervous system, even of a very simple kind, in any plant.

Scientists were not able to think of any way in which plants could be aware of human conversation. Therefore, they felt, talking to plants was silly. Most people agreed with them, until....

Some years ago, people's attitudes toward plants

PLANT MYSTERIES

Plants are as bad as people.

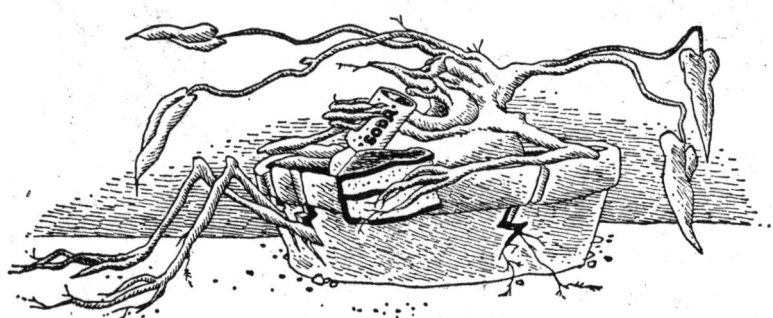

Dial·A·Plant 999·7272*

They get fat. They don't exercise. They don't eat right.

But whose fault is it?

If you're not doing what's right for your plants, chances are, they're not doing right by you.

That's where Dial-A-Plant comes in. It's a whole new service from New York Telephone featuring plant expert Jerry Baker. Each day, Jerry will give you tips on plant care along with samples of his homegrown philosophy.

So if you and your plants are ready to turn over a new leaf, call Dial-A-Plant, 999-7272.

 New York Telephone

*A one message unit charge from the 212, 516 and 914 areas. Elsewhere dial 212 first and regular toll rates apply.

Houseplants function as "pets" for many people. These people tend to fuss and worry over their plants as they would over a cat or dog. As a service to such plant owners, the New York Telephone Company set up a Dial-A-Plant line in the early 1970s.

changed. Persons who had never before said so much as "good morning" to a plant started chatting to them. As these individuals watered and tended their plants, they spoke softly to them. They urged the plants to grow, to put out new leaves, to bud.

Going beyond that, some people started playing classical music to their plants. In fact, special records of "music for plants to grow by" were sold. People also shielded their plants from loud, unpleasant noises. They even worried that their plants might find a particular place unpleasing.

Not only were plants given good care, they were also showered with good feelings.

Many plant lovers decided that plants are sensitive to people's feelings and moods. Plant lovers argued that plants know when a person admires them and when a person doesn't like them. Plant lovers contended that plants may be aware of other plants, and may even be aware of other forms of life. Plant lovers suggested that plants may be concerned for other plants' well-being, although plants may not be above feelings of jealousy and hatred.

When some of these ideas were first started, stories and jokes about conversations with plants started appearing in newspapers, magazines, radio shows, and television programs. "Do Plants Have Feelings, Too?" asked a writer for *The Baltimore Sun* in an article that later ap-

PLANT MYSTERIES

peared in *Reader's Digest*. Similar articles appeared in other papers.

Cartoonists had a field day with talking plant jokes. So did comedians on radio and television talk shows.

Much of the publicity poked fun at the idea of sensitive plants. But some of the publications took it seriously. As a result, people who had been talking to plants all along were jubilant. At last they could admit what they were doing.

"Sure, I talk to my plants," said a New York mother. "It gives them strength, sort of. I think they know I care."

"Come in and talk to our plants!" invited a big orange sign in a store window.

Several entries at a science fair contrasted a healthy plant that was the recipient of kind talk with a droopy plant that was the recipient of unkind yelling.

University professors even reported that many of their biology students believed that talking to plants might really help the plants to grow.

By the 1970s, the idea that plants may have feelings of some sort was widely held. What's more, many people believed that plants might be able to communicate their feelings.

One way a plant might signal its feelings would be by its general condition. A person might check to make sure that the plant's leaves are firm and green. Does the plant regularly put forth new stems and leaves? Does it

People Talk to Plants

blossom or bear fruit? If the answers to these questions are yes, a plant lover might consider them to be signals of the plant's good feelings. On the other hand, if a plant's leaves drop or wilt, if it drops its buds, if it doesn't grow, these indications may be signals of the plant's "unhappiness." According to some people, an extremely "unhappy" plant might signal its feelings by dying.

CHAPTER 3
But, Do Plants "Talk" Back?

MANY PEOPLE TALK to their plants, and many people believe that plants have feelings. Until recently, however, most of these people thought that plants could only communicate their feelings to us by growing well or by not growing at all.

Unlike some animal pets, plants have no tails to wag. They have no voices to bark or meow with and no ears to perk up at their owner's approach. Plants do not pant or sweat. Neither do they have hearts to pound with excitement.

Plants do, however, respire (breathe). They absorb oxygen from the atmosphere. The plant cells use it to convert food into energy. Then the waste product—carbon dioxide—is given off to the atmosphere.

Plants also have circulation systems. A plant takes in water and dissolved minerals through its roots. The resulting solution, called sap, generally circulates upward through the plant tissues.

PLANT MYSTERIES

People have asked whether a plant's respiration and sap circulation might vary in response to changes in the plant's emotions. Experts don't think so. They have measured patterns of plant respiration and sap circulation, and they never found evidence to indicate that such respiration and sap circulation were in any way affected by plant "emotions." These experts believe that plants cannot feel anything.

Therefore, if plants have feelings, there seemed, until recently, to be no way for people to learn about those feelings—other than to observe the state of a plant's health. But, signs of good or bad health can be very unreliable indicators of plant feelings because plant health can be affected by many different factors.

A plant droops. Is the cause a plant disease? A pesty bug? A shortage of moisture? Pollutants in the air? Not enough light? A lack of nutrients in the soil?

Or, does the plant droop because its feelings were hurt?

No one can be sure!

In the late 1960s some research reports hinted that there may be another way to detect a plant's feelings *if* plants have feelings. This involves measuring changes in a plant's bioelectric signals.

As in all other living things, tiny electric currents flow through each and every part of a plant. These electric currents are the movements of electrons or ions (charged

But, Do Plants "Talk" Back?

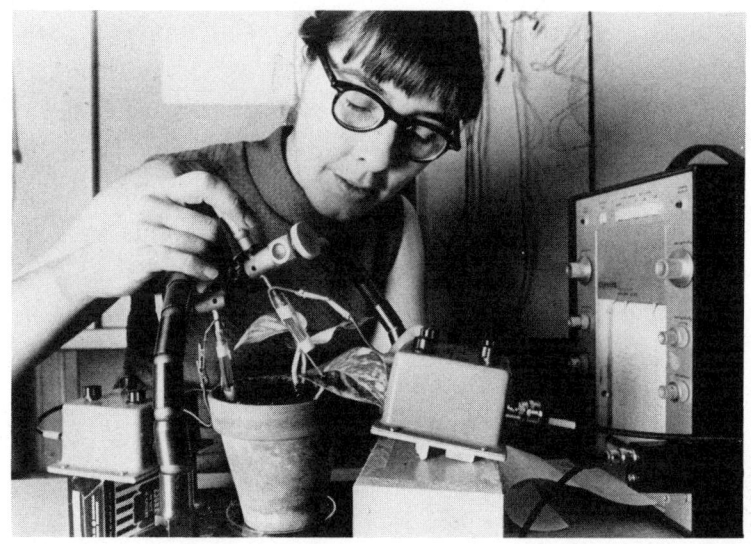

Plants' electrical activity has been recorded by many botanists including Barbara Pickard of Washington University. She stated that many of the electrical signals she recorded resembled the nervous activity of animals. (*UPI*)

atoms) from cell to cell in living tissues. Normally, this flow of electricity is much greater in animals than in plants. But it can be measured in both.

Various researchers have wondered whether the flow of electric currents in a plant might be influenced by "the state of the plant's emotions." If so, could such influences on a plant's bioelectric signals be detected and recorded? Could records of these signals be used

PLANT MYSTERIES

to prove that plants are aware of what happens around them?

Many people, including a few botanists, wondered about these things. But few attempts were made to answer these questions. Then one day a man named Cleve Backster hooked one of his office plants to a lie detector. Mr. Backster recorded some signals. The results of that informal experiment with a plant and a lie detector convinced many people that plants can indeed "talk back." Those results also set off a botanical controversy that is still not over.

Before we plunge into that controversy, let's see what a lie detector is and how it's used.

Lie detector is a commonly used name for a polygraph. A polygraph is actually a combination of instruments used by law enforcement officials in an attempt to find out whether a person is lying. This is done by detecting variations in the signals produced by certain bodily functions of the person. A typical polygraph may measure pulse rate, respiration rate, blood pressure, and galvanic skin response.

When a person lies, there's usually a nervous response. Signals representing some of the person's functions are likely to change. For instance, the person's blood pressure may rise and his pulse rate may increase.

A person being questioned is likely to be extremely nervous to start with. Therefore, the polygraph operator must be skilled enough to pick out the special nervous

But, Do Plants "Talk" Back?

responses linked to lying from various other nervous reactions. Furthermore, no two persons will react in just the same way under questioning. The nervous reactions of one person telling lies may be less pronounced than the general nervous reactions of another person who is telling the truth. A polygraph operator must be able to set up a pattern of "normal" responses for each person. Then the operator must be able to recognize variations from the "normal" when and if the person is lying.

The value of a lie-detector test depends a lot on the skill of the lie-detector operator.

Polygraph operators tend to watch for unusual responses in a few functions before they decide that a person is lying. Also, individual operators tend to pay more attention to some responses than to others. Many operators now pay particular attention to variations of galvanic skin response. That's an indication of change in the body's electrical resistance.

Change in the way a person's skin conducts tiny electric currents can indicate mental stress, nervousness, or some other emotion.

Just as all living things produce small electric currents in their body tissues, all living tissues show some resistance (opposition) to the flow of electric currents. This resistance can vary in response to changes within the living thing and to changes in the environment around the living organism.

Plants as well as animals exhibit electrical resistance.

PLANT MYSTERIES

That resistance can be measured in plants. When Cleve Backster attached his lie-detecting equipment to his office plant, he was measuring the plant's electrical resistance. For some reason, that resistance appeared to vary.

CHAPTER 4
Encounter with a Lie Detector

DURING THE LATE 1940s, Cleve Backster was instructing soldiers at Fort Holabird, Maryland, on ways to question prisoners. As part of this work, Backster received special training in lie detection, using the polygraph.

Cleve Backster was trained by experts. In time, he became a polygraph expert in his own right. Backster became particularly interested in the use of galvanic skin response (intelligence agents often call it psychogalvanic reflex) as an indication of a person's emotional state. He did much to improve the methods and equipment for measuring changes in the skin's electrical resistance. He also set up guidelines for interpreting this information.

Cleve Backster left the Army Intelligence service and joined the Central Intelligence Agency (CIA). As a CIA section head, Backster set up a polygraph examination program for the intelligence agency.

PLANT MYSTERIES

A few years later, Backster left the CIA to become director of the Leonarde Keeler Polygraph Institute. A short time after that, he founded the Backster School of Lie Detection to train state and local law enforcement officers in the use of polygraph equipment.

Mr. Backster appeared twice before a subcommittee of the U.S. House of Representatives as an expert witness on lie detection. He was twice elected to the board of directors of the American Polygraph Association.

Mr. Backster is now probably much better known for his research on plant feelings. He got involved in plant research almost by accident.

On the morning of February 2, 1966, Mr. Backster was watering his office plants. As he watched the soil absorb the moisture, he wondered if he could measure the rate at which the moisture was being transported from the roots onto the leaves of the plant. Cleve Backster decided to find out whether some of his polygraph equipment would be able to detect that movement of water through the plant tissues.

He turned on an instrument to measure galvanic skin response. Then he gently placed electrodes from the instrument on each side of a single leaf of a houseplant of the *Dracaena massangeana* species. To make sure the electrodes would stay on the leaf, Backster fastened them with a rubber band. Then he adjusted the instrument to measure the leaf's resistance to the flow of a tiny electric current from electrode to electrode. Meas-

Encounter with a Lie Detector

urements of the current flow were recorded as an ink tracing on a roll of graph paper.

When all was ready, Backster poured water into the soil around the houseplant. Then he waited for the water to circulate up into the plant's leaves.

Water containing mineral salts is a good conductor of electricity. Therefore, Cleve Backster reasoned that increased moisture in the plant's leaves and stems would reduce their resistance to electric current flow. He expected the line showing current flow on the chart to curve upward, but it didn't. The tracing on the chart briefly wavered, then it curved downward. That meant that the plant's electrical resistance had increased!

After about one minute of chart time, the tracing briefly showed an upward surge. Then it curved downward again.

Backster was mystified. But he had seen that type of tracing pattern many times before. The tracing of the changes in the plant's electrical resistance reminded him of the polygraph tracings produced by humans showing sudden emotional reactions.

Backster was aware that the galvanic skin response of a person experiencing fear will usually surge upward during a polygraph test. Polygraph operators usually consider this to be one indication that the person may be lying. A person lying would sense a threat to his well-being and thus show fear.

Backster thought about this. He wondered . . . could

PLANT MYSTERIES

the *Dracaena massangeana* be showing fear? What would happen if he were to threaten the plant's well-being?

Cleve Backster dipped one of the plant's leaves—not the one sandwiched between the electrodes—into a cup of hot coffee. Nothing happened.

Backster sat there, drinking his coffee.

Then he had another idea. Why not strike a match and actually burn the plant's leaf. As this thought passed through his mind, Backster later related, the recording pen on the instrument swept upward.

Backster had not touched the plant. He had not touched the instrument. He had hardly moved his body at all. Yet, the tracing on the chart showed a definite upward surge.

Backster wondered: Did the plant somehow know what he intended to do? Was it responding to the threat of great harm to its leaf? Could he provoke similar reactions in other plants by thinking about hurting or destroying them? Might plants have some way of knowing our thoughts and feelings?

During the next few months, Mr. Backster connected various plants to polygraph instruments. Again and again, he got tracings with his plant subjects similar to the tracings obtained with human subjects.

When threatened, the plants seemed to show "fear responses" on the charts.

Encounter with a Lie Detector

Backster repeated the experiments with detached leaves and with shreds of leaves. Each time, he reported similar results.

Mr. Backster then turned to investigating the polygraph responses of other kinds of living cells. He placed bits of fruit, vegetables, and even cells from the human body between the electrodes. In most cases, he got graph tracings similar to those he had seen in his regular detection work.

After all this, Cleve Backster started to believe that plants can perceive and react to people's thoughts and feelings. He felt that in some still unknown way plants are in touch with other living things.

Cleve Backster theorized that there exists some kind of "primary perception" in plants. He decided to design an experiment to test his ideas.

Mr. Backster wanted his experiment to be as scientific as possible. He knew that his theory would be greeted with doubt and skepticism by scientists. Therefore, he did not want to be criticized for sloppy methods or for performing unscientific tests.

Backster realized that if he used his own thoughts as an experimental threat to the plants' well-being, he could be accused of manipulating the plants in some way. He also realized that one requirement of a good scientific experiment is that it must be repeatable. That is, someone else must be able to repeat the experiment

PLANT MYSTERIES

exactly in order to check the results obtained by the first experimenter.

To achieve these conditions, Backster automated his experiment. He designed an experiment that could be run while no one was in the room. In fact, he arranged for all the "runs" or tests to take place at times when people were unlikely to be even near the room.

As the subjects for his experiment, Mr. Backster chose philodendrons. These inexpensive plants have large, firm leaves to which electrodes can easily be attached. As an "emotional stimulus" for the plants, Mr. Backster decided to use the killing of brine shrimp.

By killing the brine shrimp in front of the philodendrons, Mr. Backster hoped to get some reaction from the plants. The brine shrimp might react to the process of their deaths in a way somewhat like the emotional response of a human being. In that case, if the experimental plants were capable of "primary perception," they might pickup and respond to this reaction. Or the actual death of the brine shrimp might give off some biological signal that the plants could respond to.

Cleve Backster wanted to see if upswings in the current flow through the plants would occur at the exact moment the brine shrimp died. If that were so, then he would have evidence to back up his theory that plants can somehow perceive feelings.

Backster set up his experiment carefully. For each run of his test, he used three plants. Each plant was attached to its own instrument. In addition, Backster attached a fixed-value resistor instead of a plant leaf to a fourth recording instrument. This fourth instrument was intended to check for possible variations in the electric power supply. It would also detect other electrical disturbances that might affect the tracings of the signals from the plants.

The attached leaves of the experimental plants were first cushioned with sterile gauze and agar gel. This was meant to prevent leaf damage and also to reduce electrical resistance.

Each plant was put in a separate room, far from the area where the brine shrimp were being killed. The lighting and temperature of each room were kept uniform.

Backster arranged a system for killing the brine shrimp at random, unpredictable intervals without the presence of a human being. During any one of six possible time blocks, a machine automatically tipped a dish of brine with brine shrimp in it into a bath of simmering hot water. As they hit the hot water, the brine shrimp died almost instantly. Then the dish returned to its upright position.

A separate instrument kept track of the time blocks during which the brine shrimp were dumped into the hot water.

Cleve Backster performed seven runs of his experiment. Afraid that the plants might become used to the brine shrimp deaths and thus stop reacting to them, he

PLANT MYSTERIES

decided to use no plant more than three times. The plants and shrimp were kept separate from each other before and during the experiment.

Just before each experiment run, instruments, plants, and brine shrimp were carefully inspected. Three healthy brine shrimp were selected and placed in the dump dish. A final inspection of the graph tracings on the recording instruments followed. The automatic system for time block selection was turned on. Then, without any further human interference, the experiment run got under way.

After each run, the charts were taken out of the recording instruments and given an identification number. An experiment referee mounted the charts on cards. The referee also identified those portions of each chart which recorded the time blocks during which the shrimp killings took place.

At the same time, the chart from the recorder attached to the fixed-value resistor was examined for possible distortions caused by electromagnetic disturbances.

In addition to the regular experiment runs, seven runs were performed with sterile water containing no brine shrimp. This was done to make sure that the process of dumping the water did not by itself cause variations on the graph tracings of the recording instruments.

Cleve Backster performed seven actual experiment runs. Since three plants were used in each run, twenty-one charts were produced. Two of these charts were disqualified due to mechanical failure of the recording

Encounter with a Lie Detector

instruments. Three other charts were disqualified for showing "too much activity." And three were disqualified for showing "no apparent ability to react" during the run.

That left thirteen charts. Since each chart recorded six time blocks, that meant that there were seventy-eight bits of information to be interpreted.

Cleve Backster decided not to interpret the charts himself. Evaluating line tracings of the type coming from the plants requires expert knowledge and judgment. That judgment could be affected by an experimenter's desire to prove his theory. Backster didn't want to give anyone the chance to say that he had misinterpreted his own charts. He asked three experts to examine the charts.

Each person, working independently, interpreted the charts as he would interpret lie-detector response tracings from a human subject. The independent examiners didn't know which charts represented which runs. Neither did they know the time blocks during which the brine shrimp were killed.

The experts differed very little in their findings. The few differences that existed were easily cleared up when the experts compared their notes.

Each of the thirteen charts showed one stimulus time block—the time when brine shrimp were killed. In all, there were thirteen stimulus time blocks and sixty-five no-stimulus time blocks on the charts that were examined.

PLANT MYSTERIES

Eleven of the stimulus time blocks showed tracing reactions. They were classified as *hits*. The two stimulus time blocks that didn't show tracing reactions were classified as *misses*.

Eight of the no-stimulus time blocks showed tracing reactions and were listed as *hits*. The other fifty-seven no-stimulus time blocks contained no tracing reactions. They were classified as *misses*.

Cleve Backster was elated by these findings. In his opinion, the experiment results were evidence of "primary perception" in plants.

In the winter of 1968, Backster described his experiment in an article in *The International Journal of Parapsychology*. This journal reports "scientific" investigations of such topics as telepathy, extrasensory perception (ESP), and clairvoyance.

When he first considered his experiment findings, Backster thought he had stumbled onto a form of ESP in plants. But by the time he published his article, he had concluded that a more basic "primary perception" ability in plants was indicated by the experiment results.

CHAPTER 5
...and Related Experiments

IN THE SPRING of 1969, *National Wildlife* magazine published an article describing Cleve Backster's plant feelings experiment and findings. *Argosy* magazine followed in June 1969 with an article by Walter McGraw entitled "Plants Are Only Human."

By January 1972, even the very proper *Wall Street Journal* was paying attention to Backster's work. An article by Richard Martin entitled "Be Kind to Your Plants—You Could Cause a Violet to Shrink," appeared in the January 28, 1972, issue of the paper.

At last, the "Backster effect" was famous. That was the name given to the polygraph signal tracings from Backster's experimental plants.

Meanwhile, Cleve Backster lectured to college students and continued his experiments.

Most of Backster's later experiments seemed less carefully planned than his original brine shrimp-killing experiment. Furthermore, Backster had a disturbing way

PLANT MYSTERIES

of jumping to conclusions when it came to analyzing what the instruments attached to the plants showed.

Writers Peter Tompkins and Christopher Bird reported that Backster was once visited by a plant physiologist from Canada. She asked Backster if she could watch some of his plants while they were attached to the galvanometer. Having said she could, Backster led her to his laboratory where they started looking at graphs on the instruments attached to the plants. However, the first plant they looked at failed to respond to a stimulus. Instead, the recording pen traced a nearly straight line on the graph paper. The same thing was seen with four other plants. Only the sixth plant showed any reaction to the stimulus.

Backster was mystified. He couldn't understand why his plants had failed to react to stimuli. After a few moments he had an idea. He asked the scientist if any of her work involved hurting plants. She replied that it did. She explained that she roasted some plants in an oven to obtain their dry weight for analysis.

Backster reported that forty-five minutes after the scientist had left his laboratory, his plants once again started reacting to stimuli.

What had happened? A trained botanist would probably not even want to venture an opinion based on such a single incident. But Cleve Backster didn't hesitate to offer an explanation. He suggested that the plants some-

... and Related Experiments

how realized that the scientist's work involved killing plants. Feeling threatened by her presence, Backster said, the experimental plants fainted. Backster concluded that plants know when they are in danger and that they respond by passing out.

As Backster continued to expose plants to the killing of brine shrimp, he noted that these killings—if they took place fairly near the plants—produced less and less reaction from the plants. Again, Backster was ready to offer an explanation. He theorized that plants must possess some kind of memory. They get used to the killings going on around them. Perhaps they "conclude" that these killings are not a threat to their own well-being.

Another experiment reinforced Backster's idea that plants might be able to remember things. This experiment involved testing a plant's skill as a detective.

Backster enlisted the aid of six students from his school of lie detection. Then he put six slips of paper into a hat. Five of the slips were blank. But the sixth slip bore written instructions for someone to go into the next room and root up and destroy one of the two plants in the room. The killing of the plant was to be done in secret.

The students were blindfolded. Each one drew a slip of paper from the hat. The student who drew the slip with the instructions on it told no one about it. He quietly went into the other room and killed one of the two plants there.

The surviving plant was hooked to a galvanometer. Then, one by one, the six students walked in front of the plant. The plant showed no reaction to five of the students. But whenever the student who killed the other plant appeared, the instrument tracing showed a strong reaction from the surviving plant.

Did the plant remember the "killer?" Or was it simply picking up the person's guilt feelings? Backster wasn't sure which reason explained the plant's reaction. But he felt certain that it was one of the two.

Cleve Backster also felt that a special rapport or relationship can be established between plants and their owners. Using synchronized watches, Backster kept track of his own activities and the tracing reactions of his plants while he was away from them. He noted that the plants seemed to react to certain of his activities and feelings. In other words, his plants appeared to be "tuned in" to his thoughts.

Plants, Backster reasoned, are tuned in more to the activities and feelings of animals and humans than to the activities of other plants. He speculated that perhaps plants are more concerned about threats to their well-being from animals and humans than from other plants.

Many people who were at first skeptical of Backster's ideas and experiments began to think that he might be

... and Related Experiments

on to something. James Lincoln Collier of *The Baltimore Sun* was one of these people. Collier reported that during an interview Mr. Backster attached electrodes to a philodendron and began questioning Mr. Collier as to the date of his birth. Before the questioning, Backster told Collier to lie about his true date of birth. Then, during the questioning, Backster named seven years. Collier denied being born in any of them. However, Mr. Backster was able to name the correct year of Collier's birth solely by watching the tracings of the galvanometer attached to the philodendron. Mr. Collier left Backster's lab feeling somewhat less sure than he had been that plants can't communicate.

A psychiatrist from Rockland State Hospital (New York) and a chemist from Newark College of Engineering (New Jersey) were also shaken from their initial disbelief when they tested a human volunteer and a philodendron plant.

Dr. Aristide Esser and Douglas Dean decided to repeat one of Mr. Backster's experiments. They attached the plant to a polygraph galvanometer. Then they asked the volunteer several questions. He had been told to lie in his answers to some of the questions. The philodendron was able to pick out each lie. The results of that experiment convinced Dr. Esser that Mr. Backster's ideas about plants being able to pick up thoughts and feelings were at least partly right.

PLANT MYSTERIES

Cleve Backster wasn't able to explain how the thoughts and feelings of people and animals might reach his plants. Were the signals electromagnetic? In other words, were they radio waves, light waves, or even shorter energy waves? If so, a Faraday cage would block them. Backster put a plant in a Faraday cage. But it appeared to be able to pick up "feelings."

Might signals be similar to x-rays? If so, lead might block them. He put a plant in a lead container. But it was still able to pick up "feelings." Backster theorized that whatever the plant-animal communication link might be, it could not be any known kind of radiant energy.

Cleve Backster seems to have been interested in ESP (extrasensory perception) and other types of psychic phenomena from almost the start of his plant research. In time, he became convinced the "primary perception" abilities that plants seem to have are shared by all one-celled creatures. He tried his experiments with one-celled animals, with bits of fruit and vegetable, and with human cells. All of them seemed to have the ability to produce galvanometer tracings. All of them, Backster concluded, possess the same "primary perception" ability he believed he had found in plants.

Cleve Backster was not surprised that most scientists expressed serious doubt about his work. Though he had gone to great trouble to set up his first brine shrimp-killing experiment with plants, he did not try to repeat it.

... and Related Experiments

By the early 1970s, Mr. Backster's research seemed to be mostly with animal life. He was willing to talk to and advise people on plant experiments, and he was willing to defend the results and procedures of his own early experiments. But he seemed to have lost interest in doing more plant experiments himself.

CHAPTER 6

Leading to a "Secret Life"

AROUND THE TIME Cleve Backster was turning away from research into plant responses, the public was just learning about that research. A great surge of public interest in questions of plant sensitivity and plant awareness developed during the early 1970s.

In June 1971, *Popular Electronics* published an article entitled "More Experiments in Electroculture" by L. George Lawrence. In the article, Mr. Lawrence provided detailed diagrams and directions on how to build a "plant response detector." His design included a meter to register plant responses. Lawrence pointed out that this meter could be attached to a pen-recording device.

Lawrence also included an audio oscillator in his circuit. This vibrating device produces a sound that can be varied in pitch (frequency) by incoming electrical signals. By using this device, experimenters would be able to hear the responses of their plants. Variations in the electrical current coming from a plant would change

PLANT MYSTERIES

the frequency of the oscillator and thus change the pitch of the oscillator sound. By attaching a tape recorder to the circuit, an experimenter could record these sound changes. In that way, the experimenter would have both a sound and visual record of a plant's responses during an experiment.

Lawrence warned, however, that plants might not always respond to experiments in the same way. Echoing Cleve Backster, Mr. Lawrence declared that plants seem to have a type of consciousness or awareness located in their individual cells. He suggested that plants can sense and feel. Since plants can't escape from danger, Lawrence theorized, they might have some kind of ability to shield themselves from shock. Changes in the electrical currents within plants could resemble anxiety responses in people and other animals.

Mr. Lawrence advised his readers to treat their plants kindly. Burned or injured plants might die. Plants could also become tired out by experiments. Therefore, Lawrence wrote, it would be wise to rest the plants between experiments.

In November 1972, *Harper's Magazine* published an article that was part of a book being written at the time by Peter Tompkins and Christopher Bird. The article, "Love among the Cabbages," described many of Backster's experiments with plants. It also reported experiments of Paul Sauvin, an inventor who believed he could communicate with plants. Sauvin claimed to have used

Leading to a "Secret Life"

a philodendron to trigger a radio signal to turn on a car motor. He believed that an emotional signal sent by him to the philodendron caused the plant to trigger a radio signal that turned on the motor.

Bird and Tompkins also described Soviet research into plant responses, and mentioned other experiments that seemed to show that plants can understand and respond to various stimuli, including ghost stories.

"Love among the Cabbages" was an attention getter, but it was only a tiny foretaste of what was to come. Harper and Row brought out Bird and Tompkins' book, *The Secret Life of Plants*, in 1973. Paperback publication followed in 1974. The book reached the best-seller list. Soon everyone was talking about talking to plants.

The Secret Life of Plants is a vast collection of information on the study of plants. The authors of the book argue that plants have unrecognized abilities. They say that plants can feel and communicate, and do other surprising things. To prove their point, Bird and Tompkins describe dozens of experiments. They mention scores of experts, ranging from respected scientists to little-known mentalists.

The first chapter of the book describes Cleve Backster's research. The second, third, and fourth chapters describe the more sensational experiments of Marcel Vogel, Paul Sauvin, and L. George Lawrence.

All of these experimenters appear to be interested in one form or another of parapsychology. Vogel became

PLANT MYSTERIES

convinced that psychic energy could keep detached leaves green and healthy. He suggested that plants and people can "become one" through a sharing of the "life force." Therefore, he argued, researchers must establish empathy with their plants. In his opinion, the detached type of observation that scientists insist on just won't work with plants.

Paul Sauvin is interested in ESP and hypnotism. According to Bird and Tompkins, Sauvin often gets ideas for his inventions in psychic flashes without fully understanding the working principles of the ideas.

L. George Lawrence, an engineer and former professor of audio-visual arts, was reported to be building a biological receiving system to detect signals from deep space. This interstellar signal receiver consists of plant cells placed in a warm bath and shielded from nearby electromagnetic interference. Lawrence believes that living plant tissue is more sensitive to signals from other worlds than electronic detectors. In fact, Lawrence believes that his device may have already picked up signals from space. Lawrence believes in the existence of a "paranormal matrix." He describes this as a unique communications grid which binds all life together. He says that the grid operates beyond currently known physical laws. This unusual research is sponsored by the Anchor College of Truth, of which Lawrence is a vice president.

In addition to descriptions of the efforts of Backster,

Leading to a "Secret Life"

Vogel, Sauvin, and Lawrence, *The Secret Life of Plants* presents some Soviet research into plant responses and describes the work of some famous plant scientists. These include Sir Jagadis Chandra Bose (1858–1937), an Indian physicist and plant physiologist, Charles Darwin (1809–1882), Luther Burbank (1849–1926), and George Washington Carver (1864–1943).

The authors then discuss research on plant responses to sound and to other radiations, and describe experiments and projects that appear to demonstrate the superiority of organic farming. Bird and Tompkins oppose the use of chemical fertilizers, contending that such fertilizers poison our soil and food. They cite numerous experiments in nutrition and in soil restoration to support their argument.

Many of the authors' observations make good sense. Some of the research they describe appears to have a scientific basis. However, Bird and Tompkins do not appear to have been selective about the plant research they describe in their book. Neither were they selective about the men and women they quote in support of their ideas. Some of the experiments described in *The Secret Life of Plants* were performed by biologists and physicists working in their own fields, but many of the experiments were not. Several of the biological experiments were performed by experts in some scientific field other than biology.

Marcel Vogel is one example. Appearing as an expert

PLANT MYSTERIES

on a television show, Mr. Vogel displayed tracings to demonstrate a plant's reactions to his thoughts and feelings. However, Mr. Vogel had no formal training in plant biology. He studied magnetics and the behavior of liquid crystals, and his inventions contributed to the advance of computer technology.

L. George Lawrence and Paul Sauvin are also experts in fields other than plant biology. Mr. Lawrence is an engineer. Mr. Sauvin is described in the book as an electronics specialist. By training and experience, it is likely that these men understand how to record the electrical activity of plants, but it seems less likely that they fully understand how to interpret the meaning of the information they collect.

Some of the research described by Bird and Tompkins in the book may not have been the work of scientists. In chapter 17, the authors suggest that plants may be able to transmute one chemical element to another. This changing of elements was the old dream of the alchemists. But as far as modern chemists know, this is not possible by ordinary chemical means although nuclear scientists have achieved it by bombarding atomic nuclei.

A number of agricultural experiments described by Bird and Tompkins hint that plants might be able to change one nutrient element into another. Many of these experiments were performed in the last century and it's likely that not enough care was taken to insure that the plants were not getting the new elements from their

Leading to a "Secret Life"

environments in some unrecognized way. Strictly speaking, however, there is nothing unscientific about the idea of plants changing elements. It is just extremely unlikely. Someday, some scientists might show that this happens.

Bird and Tompkins were not satisfied with this. They quote a book entitled *The Nature of Substance* by Rudolph Hauschka. In this book, Hauschka calls the chemical elements "corpses"—the remains of life forms. He suggests that life existed before matter and came out of a spiritual cosmos. Plants, according to Hauschka, can generate matter from a nonmaterial region and make it disappear again.

This suggestion bewilders most scientists. Although no one can prove with absolute certainty that it is impossible, there is no evidence to support such an idea. This notion can be described as a mystical idea. But it is not, in any sense, scientific.

Unfortunately, many of the suggestions and experiments described by Bird and Tompkins in *The Secret Life of Plants* are based on similar mystical ideas. In many cases, some force beyond our current knowledge or outside physical laws as we know them is said to be responsible for the result described or for the observation recorded. Cosmic principles, mysterious forms of energy, or radiation from unknown sources are used to explain many of the unusual discoveries mentioned in *The Secret Life of Plants*.

To the scientist, such broad "explanations" are not

PLANT MYSTERIES

explanations at all because they don't really tell us anything. They do not show us how or why the unusual results appeared. Scientists often suspect that some simpler and less unlikely theory would explain the results of the experiments better. In other words, scientists prefer the simplest explanation.

Furthermore, scientists criticize many of the methods of researchers discussed in *The Secret Life of Plants*. They point out that some of the ways the researchers measured results were not accurate. Perhaps coincidence was responsible for many of the results reported. Perhaps many of the results simply cannot be explained. Scientists are used to living with uncertainty. Life scientists in general concluded that much of the material in *The Secret Life of Plants* was "nonsense."

Why then was the book so popular? Why did so many readers take the basic idea that plants have a secret emotional life so seriously? There are many answers to those questions.

The book came at the right time. Many people were upset by the abuse of our environment. They worried about the traces of chemical pesticides in our food. They disliked the "unnaturalness" of much of modern life. Some of these people found support for their ideas in *The Secret Life of Plants*. Much of the book's criticism of chemical farming may be justified. Several of the soil experiments reported in the book seem well done and convincing.

Leading to a "Secret Life"

Many new discoveries are now being made by plant scientists. People are beginning to realize that the plant kingdom is full of surprises. Plant biologists are finding that plants have certain similarities to animals. Plants definitely exhibit electrical responses of various kinds. The ways in which they respond to light are complex. Scientists are just beginning to understand these complex responses. Scientists are also finding that plants may respond, by growing faster, to certain kinds of sound.

Recent research has shown us that the subject of how plants react to their environments is full of fascinating surprises. Plants appear to be more sensitive than many scientists had previously thought them to be. Much of this recent research is presented in *The Secret Life of Plants*. That is surely part of the book's appeal.

But there's more to it than that. It is exciting to believe that plants are in some way like us. It is even more exciting to think that plants care what we think of them. Many people believed that anyway. They cared for their plants and they could see the plants respond to that care. They talked to their plants and they half believed that somehow the plants understood what was being said to them.

The Secret Life of Plants confirmed and made respectable thoughts that many, many people had held all along. It's no wonder the book became a best seller.

CHAPTER 7
Do Plants Have Feelings?

IN 1973, one of the people who considered that plants may have feelings was a fourteen-year-old high school student, Ursula Schwebs. Ursula read "Love among the Cabbages," in *Harper's Magazine*. Bird and Tompkins' description of Cleve Backster's experiments impressed her. Ursula interviewed friends and neighbors, and she found that several of them talked to their plants. Finally, Ursula decided to repeat Cleve Backster's experiments for her school's science fair.

Ursula did not have a lie detector. But she knew that the portion of lie detector equipment used by Backster in his plant experiments normally measures the electrical resistance of a person's skin. Ursula borrowed her father's electronic volt-ohmmeter. This instrument can measure both voltage and resistance. It is probably much more sensitive than Cleve Backster's galvanometer.

Ursula also borrowed poinsettias, African violets, and philodendrons from her mother. These plants were to be her experimental subjects.

PLANT MYSTERIES

Ursula then rounded up plastic bags and a vaporizer. Using paper clips and office binding material, she made contacts to attach the plants to her ohmmeter.

At last, Ursula Schwebs was ready to repeat Cleve Backster's experiments on plant responses.

Ursula hooked up a poinsettia to her volt-ohmmeter. The instrument immediately showed resistance and voltage readings. Ursula did not jump to the conclusion that the readings on the instrument were caused by a deliberate response from the plant. She had expected the plant to have resistance. But she was puzzled by the voltage reading. She was particularly mystified when she found that the voltage reading increased slightly whenever she walked toward the plant.

Ursula wondered . . . why did the meter indicate a slight direct current (DC) voltage? Why did that slight voltage seem to increase as she approached the plant even though she didn't touch the plant? What exactly was she measuring? Ursula decided to find out.

Taking the wires from the volt-ohmmeter into her hands, Ursula measured her own body resistance. At the same time, she found that her body generated some voltage. Later, Ursula also discovered that the bodies of her friends generated voltages. But these voltage readings were not the same every day.

Ursula soon realized that everyone carries an electrical charge on his or her body. She also found that this charge varies with the moisture content of the air around

Do Plants Have Feelings?

the person. When the air is dry, the electrical charge on a person is relatively high. She also found that the charge can be affected by the shoes a person wears. If a person wears leather-soled shoes, the built-up electric charge on the person will slowly discharge to the ground. Rubber-soled shoes tend to prevent this.

Ursula concluded that plants, like people, tend to build up an electric charge, and this charge is affected by conditions of the atmosphere around the plant. Ursula concluded that she could change the charge on a plant just by walking toward it—she did not have to touch the plant.

Having made these discoveries, Ursula Schwebs decided to go on with her attempts to repeat Cleve Backster's experiments.

Ursula attached her volt-ohmmeter to an African violet. The meter indicated a resistance reading of 5.3 million ohms. Then she watered the violet. As she poured in the water, the instrument's needle moved. Ursula concluded that the needle moved because she had moved toward the plant to water it.

When Ursula moved away from the plant, the ohmmeter needle remained still for about ten minutes. Then it climbed to a reading of 6.3 million ohms. That was an increase in resistance of about 10 percent. The voltage increased by a similar percentage.

Backster had reported that the resistance of plants increased when they were watered. He had initially ex-

PLANT MYSTERIES

pected the plants' resistance to decrease. In this case, Ursula's findings matched Cleve Backster's findings.

Did that mean that the African violet was showing pleasure at being watered? Were the increases in resistance and voltage the plant's way of expressing its feelings?

Ursula was skeptical.

"I don't know if my plant was stimulated or not," she said. "The short wiggle was the same if I poured water or not. The long reaction to watering could be explained by the plant and the water together taking on a larger charge than before."

Ursula next decided to test Backster's brine shrimp-killing experiment. However, Ursula did not use brine shrimp in her experiment.

She arranged a plastic bag over a plant to hold in moisture in the air around the plant. Then she poured boiling water next to the plant. Within twenty minutes, the plant's electrical resistance dropped from six million ohms to two million ohms.

Why? Ursula hadn't killed anything. She reasoned that the drop in electrical resistance could not be explained by the plant picking up a signal from a dying organism.

Ursula noted, however, that the humidity of the air around the plant increased when she poured the boiling water. As the air became increasingly moist, the plant's electrical resistance decreased. This, she realized, might

Do Plants Have Feelings?

explain the drop in a plant's resistance during Backster's shrimp-killing experiments. If the brine shrimp were dumped into boiling water near the plant, steam from the boiling water would increase the humidity of the air around the plant. As a result, the plant's electrical resistance would decrease.

This might also account for another of Backster's findings. Backster reported that his plants seemed to get used to the brine shrimp killings. After some time, the plants would no longer respond to the killings. Ursula reasoned that this could be the result of the air around the plants becoming saturated with moisture from the boiling water used for shrimp killing. In that case, the resistance readings of the plants would be as low as they could go. So, the plants could show no further responses.

Ursula then decided to try Backster's original "threat to well-being" experiment. She lit a candle and walked toward a wired-up poinsettia. The needle of the ohmmeter attached to the poinsettia quivered a little as she approached the plant. Ursula then held the candle under a poinsettia leaf. But there was almost no reaction on the meter.

Ursula did, however, get a changed reading on the meter when she moved her chair. She also got reactions when she approached the plant with a match. It made no difference whether the match was lit or not.

Ursula Schwebs concluded that the volt-ohmmeter was really reacting to her electric charge.

PLANT MYSTERIES

Ursula was also able to find a possible explanation for Backster's report that his plants fainted in the presence of a physiologist who "killed" plants. Ursula took turns walking toward her plants while wearing different pairs of shoes. She found that she obtained almost no reading on the meter when she approached the plant on a humid day while wearing rubber-soled shoes. If she wore leather-soled shoes, the plant would react. But if her shoes were wet, the plant would not react.

Did the plant physiologist from Canada wear rubber-soled shoes when she visited Backster's lab. Had it been a rainy day?

Ursula didn't know. But she considered that she had provided a reasonable alternative explanation for the lack of response from Backster's plants when the "plant killer" was with him.

Ursula spent a total of 172 hours working on her science fair project. She won an award for it. The science fair judge suggested that Ursula send a description of her project to *Harper's Magazine*, since that magazine had published "Love among the Cabbages" by Bird and Tompkins.

Harper's Magazine accepted Ursula's article and published it in their June 1973 issue.

In the following year, Ursula entered Washington-Lee High School in Arlington, Virginia, where she continued her experiments with plants and electricity. She found that plants from dry environments are more re-

Do Plants Have Feelings?

sponsive to watering than plants from wet environments. All her plants showed increased voltage and resistance readings after they were watered, but the effect was more marked with plants from dry environments.

Ursula also found that light conditions can change the electric potentials of plants.

By applying electric currents from various sources to her plants, Ursula discovered that the plants' electric potentials (voltages) could be increased in this way. When the applied current was removed, she found that it took about an hour for the plants to return to their original voltage levels.

Based on this work, Ursula wrote an article that appeared in *Chemistry,* May 1974, in which she concluded that plants generate small amounts of electricity on their own. She also concluded that plants' electric potentials are influenced by electricity in their immediate environment. These facts, she believes, explain the measurement on which claims that plants have feelings are based.

"According to my experiments," she wrote, "there is no evidence that plants have feelings as we human beings know them."

Had Ursula actually shown that Cleve Backster's experimental results were simply due to changes in the environmental electricity around his plants?

Not completely! Ursula's experiments and the way she interpreted them were logical. Her results seem to ex-

PLANT MYSTERIES

plain Cleve Backster's statements as they were reported in the article, "Love among the Cabbages," in *Harper's Magazine*. However, that article did not describe Backster's shrimp-killing experiment in detail.

From her articles, it appeared that Ursula had not read Backster's original account of his experiment in *The International Journal of Parapsychology*. She was not aware that the brine shrimp killings in Backster's main experiment were carried out in a room far from the place where the plants were monitored. Furthermore, Ursula did not repeat Backster's experiment in the way that he did it. She did not kill shrimp. She did not say how many times she poured boiling water near plants. She did not say how many plants she used. Ursula did not really disprove Backster's conclusions from the shrimp-killing experiment.

But Ursula's work did raise doubts about many of Backster's statements. She was very careful to find out exactly what she was measuring with her ohmmeter. Her findings on the effects of her own movements on the electrical responses of plants made people wonder whether some of Backster's results could be explained by his movements around his plants.

In his original experiments Cleve Backster considered and allowed for interference from other electrical devices when interpreting his results. But he didn't mention the possibility that his own movements might change the readings of electrical resistance from his plants.

Do Plants Have Feelings?

Did Backster assume that the effect of his movements would be too small to make a difference in the readings? Or was he sure that he never moved in a way that would change his plants' responses? Did he ignore static electricity?

We have no answers to these questions. But the questions remain for any consideration of Cleve Backster's work and of Ursula Schweb's work.

Ursula's work provides possible explanations for some of the happenings reported by Backster. More important, however, Ursula's approach and conclusions give us an excellent example of the way in which scientists think. Scientists always look for the simplest explanation for an event that is consistent with the facts.

When Backster saw the unusual tracings of responses from his plants he jumped to an unusual conclusion. He guessed that the plants were experiencing feelings based on the similarity of the plant tracings to tracings produced by people he tested. This idea may have influenced all his later experiments and conclusions.

Scientists get ideas that way sometimes. They call it a hunch or intuition. But they are usually very critical of those ideas. Before they test such notions, they try to think of an easier way to explain their results. Then if the original hunch still seems reasonable, they test it very thoroughly.

Normally, scientists do not work by intuition. They form a hypothesis or theory of why or how a thing hap-

PLANT MYSTERIES

pens based on observation, then they carefully plan an experiment to test the theory. Such an experiment is so planned that there will be only one possible reason for the result—if the experiment works. Scientists always want to be sure of what they are measuring, and what their results really mean. They look first for the most likely explanation for the events they have observed. Only if that explanation fails to account for their results do they try a more unusual explanation.

Scientists almost never propose such general, vague ideas as "primary perception" or "extrasensory perception" to explain experimental results. Scientists like their explanations to be as clear and as specific as possible. They would rather say that they don't know why something happens than try to explain it in vague, general terms.

No one, including Backster, can say precisely what "primary perception" is. Neither can anyone say what "extrasensory perception" (ESP) really is. People who believe in ESP talk about unknown sources of energy or undiscovered radiations. But these terms don't really tell us anything; therefore, scientists do not find such ideas useful.

Ursula Schwebs put it this way: "Backster and others are working on the idea of ESP working with plants. I believe they were unable to understand some of their experiments, so they used the idea of ESP to explain them."

CHAPTER 8

The Experts Take a Turn

URSULA SCHWEBS WAS NOT the only person to test Cleve Backster's experiments. At least two groups of scientists, working independently, tried to repeat some of Backster's tests of plant sensitivity.

Most plant biologists doubted Backster's claim that plants can feel and communicate, but much of the general public believed that claim. So, a few scientists felt challenged to test Backster's claim.

Furthermore, Backster had conducted an experiment and published a report of it to support his theory. Some scientists realized that it would not be scientifically proper to ignore that experiment. When a scientist publishes a report of an experiment to support a new theory, he expects other scientists to repeat his experiment. In a way, he has a right to expect that someone else will test his results.

Many plant biologists did not consider Cleve Back-

PLANT MYSTERIES

ter's theory and experiment to be worthy of such testing. They did not view Backster's work as serious or scientific. However, some biologists remembered that many of what are now respected science theories were once considered "scatterbrained notions." At least two attempts were made by scientists to repeat Backster's brine shrimp-killing experiment.

Professor Edgar L. Gasteiger of Cornell University was in charge of one of these attempts. Dr. Gasteiger and his students studied the way Backster had conducted his experiment. Before attempting to repeat the experiment, they talked to Backster. They noted the precautions he took to shield his experiment from outside interference, and they added a few precautions of their own.

On the day of the experiment, the Cornell scientists placed the plants in Faraday cages to shield them from electromagnetic interference. They checked the recording system to make sure it was working properly. Then, as they hooked the plants to the lie detector equipment, they were careful not to hurt the leaves in any way. The scientists also grounded the plants to further reduce the chances of electromagnetic interference.

Just before the shrimp-killing runs, the scientists examined each brine shrimp under a microscope to make sure that it was alive and healthy.

During the automated shrimp-killing runs, the experimenters stayed away from the test area. In this way, they

The Experts Take a Turn

considered that none of their thoughts or impulses could influence the test results.

When the runs were completed, Gasteiger and his students tallied the results. They had to disqualify three of their eight sessions because of recording failures. That left them with a total of sixty brine shrimp-dumping periods. (Backster's experiment had given only thirteen such periods.) Only ten positive reactions, *hits*, were recorded during those sixty periods. Six hits were recorded during forty periods when the apparatus dumped sterile water instead of brine shrimp.

Professor Gasteiger and his students studied their results very carefully. Using statistical methods, they concluded that the ten hits could be accounted for by chance or coincidence. In scientific terms, the number of hits was statistically insignificant.

Professor Gasteiger and his students had repeated Backster's experiment, but they had not achieved Backter's results. They considered that they had not confirmed Backster's theory.

Dr. John M. Kmetz of Science Unlimited Research in San Antonio, Texas, also repeated Backster's experiment.

Like Professor Gasteiger, Dr. Kmetz first consulted Mr. Backster about setting up the tests. He followed Backster's precautions.

Until the time of the actual test, Dr. Kmetz kept the plants and brine shrimp apart. At Backster's suggestion,

PLANT MYSTERIES

he used only mating pairs of brine shrimp. Dr. Kmetz also avoided being in the laboratory while the test was taking place.

Dr. Kmetz ran two different series of shrimp-killing experiments. In the first series, he used an instrument to measure galvanic skin response, just as Backster had done. In fact, Kmetz's instrument was made by the company that made Backster's instrument. Except for the pen-centering devices on them, the two instruments were similar.

In the first series of tests, Dr. Kmetz used 42 philodendrons as test subjects. He recorded 252 shrimp-killing periods. He recorded 252 periods during which sterile water instead of shrimp was dumped. Kmetz noted that it made no substantial difference in the number of hits whether brine shrimp or sterile water was dumped.

Mr. Backster argued with Dr. Kmetz over the results of the first series of tests. Mr. Backster contended that the results might have been affected by the use of a different pen-centering device on Dr. Kmetz's recording instrument.

Dr. Kmetz ran a second series of tests. This time, he used a different response-recording method—one that measured electric potentials on the plants' leaves. In the second series of tests, Dr. Kmetz used twenty-one plants. He reported once again that the plants failed to react in any meaningful way to the killing of brine shrimp.

The Experts Take a Turn

This raised a nagging question. Why did Mr. Backster get positive reaction results (hits) in his shrimp-killing experiment when Professor Gasteiger and Dr. Kmetz failed to get such results in their repeats of the experiment? Dr. Kmetz concluded that Mr. Backster's positive reaction results were the "combined results of faulty technique and equipment."

Cleve Backster, on the other hand, blamed the failure of the scientists to match his results partly on small differences in the instruments they used. He also contended that the spirit in which the two scientists conducted the experiment was "not right." Mr. Backster implied that the scientists could not repeat his experimental results because they were not in proper harmony with their plants. He also accused Dr. Kmetz of forcing a delicate experiment.

Talking to a reporter for *The Christian Science Monitor*, Cleve Backster stated that it is very hard to plan repeatable experiments on plants. "Mother Nature," he said, "doesn't want to jump through a hoop ten times in a row simply because someone wants her to."

Members of the scientific community took a dim view of that response. As far as they were concerned, experimental findings must be repeatable in order to be considered valid. In their opinion, Backster's ideas had had a fair trial—and failed.

But that was not the end of it.

A noted plant biologist, Professor Arthur W. Galston

PLANT MYSTERIES

of Yale University, organized a session on the topic for the 1975 meeting of the American Association for the Advancement of Science (AAAS). The session was entitled: "Electrical Responses of Plants to External Stimuli."

Cleve Backster was invited to the session. He was asked to present his theory and to describe his experiments. Professor Gasteiger and Dr. Kmetz were invited to discuss their attempts to repeat Backster's shrimp-killing experiment.

During the session, Cleve Backster acknowledged that he had not proven that plants possess primary perception. He also stated that he had never repeated the shrimp-killing experiment. However, he maintained that his original shrimp-killing experiment was not a failure.

Backster went on to explain that he had stopped killing living things. He suggested that other people should stop dropping brine shrimp or other living things into boiling water. Cleve Backster then told a crowded roomful of AAAS members that he was experimenting with bacteria in yogurt. In those experiments, he said, he was looking for positive responses instead of trying for reactions to cell death. Backster described an experiment in which he fed milk to bacteria in a container of yogurt and obtained an electrical response from a different container of yogurt.

Dr. Kmetz then reported that he had tried and failed to duplicate the yogurt experiment.

The Experts Take a Turn

Cleve Backster continued to maintain that living organisms, such as plants and yogurt bacteria, can sense and feel.

The scientists, in effect, threw up their hands and left. Why, some reporters asked, had the AAAS scheduled such a session to begin with?

In reply, Professor Galston explained that he was worried because so many people believed that plants could communicate with them. He considered this to be incorrect. He organized the AAAS session to set the record straight.

Professor Galston had taken an informal survey of the students in his biology class at Yale, and he found that half of his students talked to their plants. What's more, half of that group really thought it did some good. The professor was upset. Such an idea is not only unscientific, he reasoned, it might be dangerous. In a world beset by problems of too many people and too little food, the belief that talking to plants will make them grow better might hinder agricultural progress.

Before the 1975 AAAS meeting, many sensible people believed in a type of plant awareness. Most plant scientists considered this belief to be unfounded, unscientific, and unreasonable. Professor Galston considered that scientists had a responsibility to present their views to the public. That led to the session on "Electrical Responses of Plants to External Stimuli."

From Professor Galston's viewpoint, the session was

PLANT MYSTERIES

probably a success. Despite Mr. Backster's refusal to back down, the professor seemed satisfied that Backster's work had been shown to be unscientific. If plants have any kind of consciousness, no one had demonstrated its existence in a way that satisfied plant scientists.

However, many questions were left unanswered. If plants are not in any way conscious, how do they "know" so well when to bud and flower? How do they "know" how to find light, water, and other life-support aids? What makes *Mimosa pudica* fold and droop its leaves at the slightest touch? What makes the Venus's-flytrap snap its bristled leaves shut on an unwary prey?

How can plants respond so precisely to their environment if they are unaware of that environment? Are the higher life forms of plant life less aware of their surroundings than the lower forms of animal life? In this sense, how are plants and animals similar? And how do they differ?

CHAPTER 9
A Certain Unity of Life

PLANTS AND ANIMALS have a lot in common. All life, according to the theory of evolution, had its origin in the sea. Somewhere in the dim, distant reaches of time, plants and animals had a common ancestor on this water planet.

The chemical makeup of plant and animal life is fundamentally the same. All life is composed of complex carbon compounds, called organic compounds. Plants and animals have bodies composed of one or many cells. These cells have many structural similarities.

Within their bodies, plants and animals must maintain conditions that are very different from the conditions in their immediate surroundings. To do this, plants and animals must expend energy. They must acquire or manufacture food. They must exchange gases with their environment. They must absorb and retain sufficient moisture for life processes. And to continue their species, plants and animals must be able to reproduce their kind.

PLANT MYSTERIES

In short, individual plants and animals must adjust to their environment.

Animals adjust to their surroundings in ways that are more obvious to people than are the ways in which plants adjust. Most animals are able to move from place to place. They can flee danger and they can seek out a more advantageous environment. Plants, on the other hand, are usually rooted to one spot.

There are exceptions to this. Some marine animals permanently attach themselves to the sea floor, while some forms of seaweed are free-floating. Certain plants, such as witch hazel, garden beans, and peas can send their seeds "traveling." These seeds are sometimes shot out over considerable distances from the parent plants. Other plants send out shoots from roots or leaves to spread over large areas or to travel to new parts of a forest or garden.

Though most plants are fixed to a permanent location, they are able to move their bodies in various ways. The most important of these plant body movements are growth movements called tropisms.

A tropism is the growth movement of a plant in response to an environmental stimulus coming most strongly from one direction. Such a movement toward the stimulus is called a positive tropism. If the growth movement is away from the stimulus, it is called a negative tropism.

Most plants will, for instance, grow toward a light

A Certain Unity of Life

source. Houseplants growing in a window box or inside a window will bend toward the light. If they are not rotated frequently, these plants will grow lopsided. This growth of a plant toward a source of light is called phototropism.

Some plants, however, prefer low-light conditions. When one of these plants, such as the coleus, is exposed to very bright light, it will tend to bend away from the light source. This is called negative phototropism.

Certain plants actually seem to search for darkness. This growth movement toward darkness is called skototropism. It was first suggested in 1975 by two scientists who normally study insects. While walking through a tropical forest in Costa Rica, Dr. Donald R. Strong, Jr., and Mr. Thomas S. Ray, Jr., noticed that vine seedlings on the forest floor were always pointed toward a nearby tree. Like the spokes of a wheel going in toward the hub, the snakelike vines converged on the nearest tree trunk. It was as if the vines "knew" where the nearest tree was.

The two scientists studied this strange behavior. They conducted several experiments with vine seedlings and dark objects in the forest and the laboratory. They concluded that certain vines have evolved with the ability to "sense" the location of the darkest nearby object, which in most cases would be a tree for the vine to climb.

In locating and growing toward the dark object, the response of the vine is skototropic. But when the vine reaches the tree and starts to climb, its growth movement

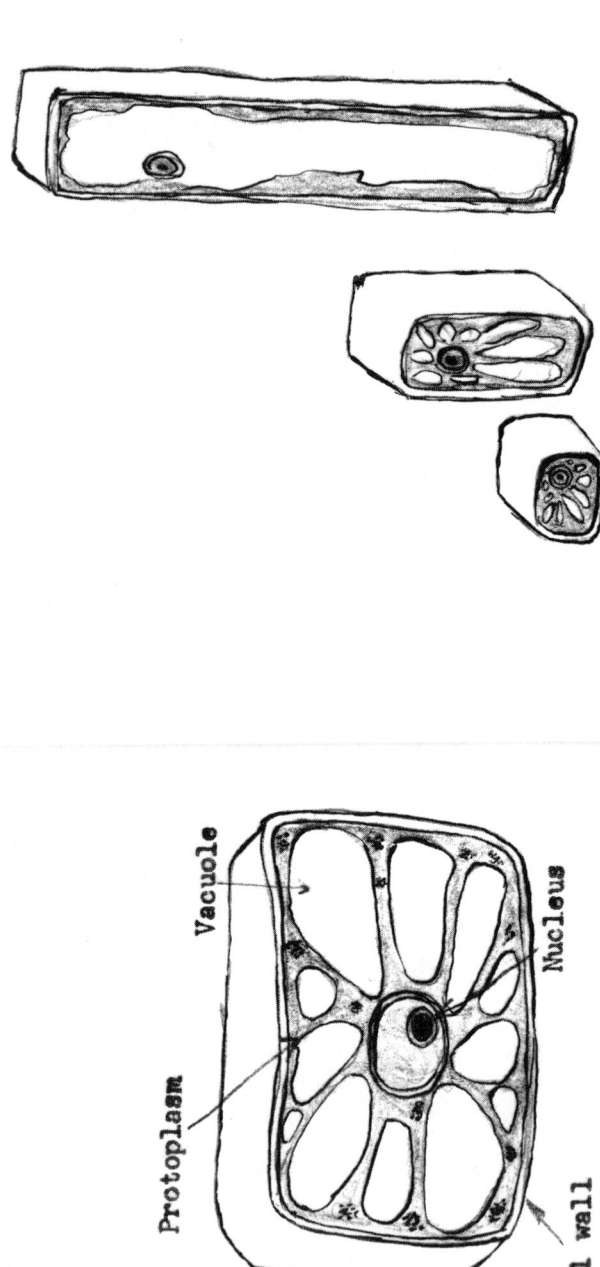

PLANT CELL

CELL ENLARGEMENT

All plant cells grow by a process of enlargement in a predetermined direction. The amount of protoplasm in each cell remains the same, but the vacuoles take in water and swell. This stretches the protoplasm and cell walls. When a cell becomes relatively large, the many original vacuoles combine to form a single water-filled vacuole.

A Certain Unity of Life

changes from skototropic to mostly phototropic. In other words, the vine climbs toward the light.

There are other tropisms and some movements that are not considered tropisms involved in the growth of the climbing vine.

Vines, like most other plants, also show positive and negative geotropic responses. Plant roots usually grow downward in response to the force of gravity. This is called positive geotropism. Plant stems show an opposite tendency when they grow upward away from the pull of gravity, which is called negative geotropism.

Roots also tend to grow toward a source of water, provided that there is enough moisture in the soil. This is called hydrotropism.

Certain plants also respond to contact with solid objects—tree trunks, branches, wires, walls, poles, trellises—by climbing up those objects or by curling around those objects. Any plant response to touch or contact is known as a thigmotropism.

A particular type of growth movement called nutation also enables climbing plants to support themselves on certain objects. In nutation, the tips of the plants trace spiral paths through the air as they grow. The tips of most nutating plants spiral in a counterclockwise direction. But a few of them move clockwise. Nutations are caused by the unequal growth of plants' stems.

When the entire growing organ of a plant is affected uniformly by an environmental stimulus, the resulting

As plants grow, their tips move back and forth. The motions of these young sprouts are almost snake-like. Scientists call these wiggling plant movements nutations. Charles Darwin suggested that such movement is an innate or inherited characteristic of plants. Other botanists have since suggested that the movement is a plant response to gravity. In 1980, a space experiment may provide a definite answer. Twenty sunflower plants will travel in a space vehicle beyond Earth's gravitational field. A time-lapse camera system will monitor the plants' growth, and the information will be transmitted to Earth. If the plants stop wiggling as they grow, that will establish that the nutations are in response to gravity. If the wiggling motion continues, that will establish that nutation is in their genes. (*Union Carbide Corporation*)

A Certain Unity of Life

response is called a nastic movement. Changes in diffuse light and variations in temperature can trigger nastic movements in flowers such as morning glories and moonflowers, which open and close at specific times of the day in response to light changes. The leaves of some plants fold up or change their positions at night. These sleeping patterns of the leaves are also nastic movements. Certain flowers remain closed on cold days and open up on warm days. Rhododendron leaves droop and curl in cold weather, but they spread out in warm weather. All these responses to environment are classed as nastic movements.

Tropisms, nutations, and nastic movements are relatively slow. However, some plants are also capable of rapid body movements. A Venus's-flytrap's leaf will suddenly snap shut to trap a fly. The leaves of *Mimosa pudica* will suddenly furl and collapse in response to touch, or sound, or shock.

Such spectacular movements of plants are caused by sudden increases or decreases in the volumes of plant cells. These sudden changes in cell volume usually result from rapid inward or outward flow of water in the plant tissues. Movements of this type are known as turgor movements.

As we've seen, plants move in various complex ways. They also respond in complex ways to changes in their environments. Temperature, light, and the presence or absence of water can all affect the germination of seeds and the growth, movement, and flowering of plants.

CHAPTER 10
Responses and Adaptations

PLANTS NATIVE TO temperate climate zones are particularly sensitive to temperature and light changes. That's probably because these plants have had to adapt themselves to periods of unfavorable conditions.

The seeds of many of these plants go through periods of dormancy. Such a seed will not sprout during the period of dormancy even though conditions for sprouting may be favorable.

Length of seed dormancy and conditions to trigger an end to dormancy vary from plant species to plant species. In order to sprout, most seeds require a period of rest. Some seeds require chilling as well. Some seeds need one or more exposures to light after the chilling so that the seeds will sprout.

The growth and flowering of plants depend on the amount of daylight available to those plants. Dormancy is triggered in many plant species by the shortening of the days as fall approaches. Some of these species need only the return of long days to start growing again. Others, such as maple and apple trees, need a spell of cold weather as well.

Plants can be placed in three categories, according to the conditions under which they will flower. The categories are short-day, day-neutral, and long-day.

Short-day plants flower in winter. These plants need a long period of uninterrupted darkness in order to flower. Laboratory tests have shown that even a brief exposure to light will prevent them from flowering.

Long-day plants flower during the summer. Their flowering response is triggered by short periods of darkness.

For day-neutral plants, however, the length of the period of darkness seems to make no difference.

Plants that key their growth and flowering to light conditions are said to show photoperiodism. Some animals, particularly those that hibernate, also show a type of photoperiodic response. But, in general, animals seem to be less dependent on light and temperature conditions than are most plants.

Plants and animals also show other overlapping patterns of similarities and differences. Most plants manufacture their food by the process of photosynthesis, while animals eat plants or other animals. However, some plants also appear to eat animals.

In the process of photosynthesis, plants use light energy to change carbon dioxide, water, nitrogen, and

Responses and Adaptations

various traces of mineral nutrients into usable carbon compounds. The plants then use these compounds to build their bodies and to carry on metabolic processes.

Plants take in water and dissolved minerals through their roots. Most plants get sufficient nutrients in this way. But some plants grow in soil that is very poor in mineral nutrients needed by those plants. How do those plants cope with that situation?

The plants have adapted to life in those areas by developing an animal-like method of supplementing their diets. The plants get needed nutrients by catching, consuming, and digesting small animals!

Carnivorous plants fascinated Charles Darwin—he spent two years experimenting with them. He and his son Francis set up two groups of similar carnivorous plants. They placed one group under a glass cover to keep those plants from catching insects. They regularly fed meat to the plants in the other group. The plants that were deprived of meat did not die, but the plants that were fed meat grew much taller and stronger. This showed that carnivorous plants get extra needed nutrients from the insects they consume and digest.

However, long after that experiment, some scientists continued to voice doubts that carnivorous plants actually digest and assimilate the insects they catch.

It wasn't until the ability to label substances radioactively was developed that scientists were able to prove that such digestion takes place.

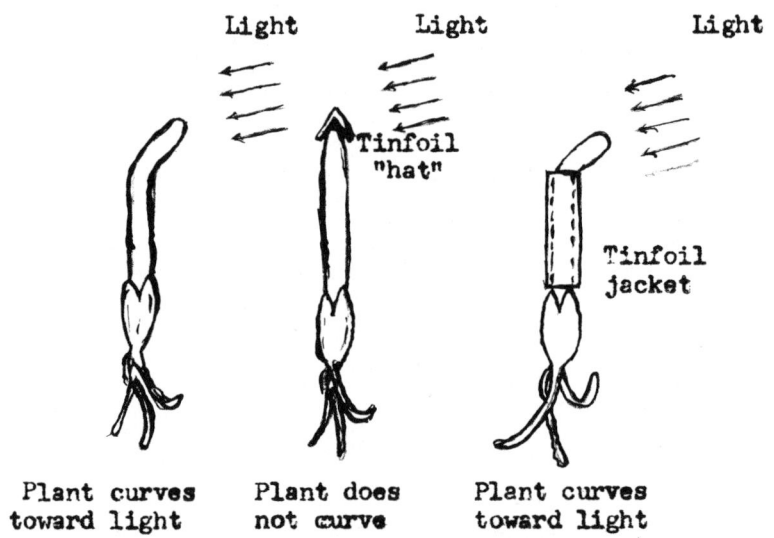

Charles Darwin and his son Francis investigated the phenomenon of phototropism in 1881. They noted that the tip of a growing coleoptile—a hollow sheath of grass—ordinarily curves toward a light source (left). The Darwins prevented this bending toward light tendency by placing an opaque (tinfoil) cap on a coleoptile tip (center). Later, they placed a tinfoil jacket on a coleoptile stalk and left the tip uncovered (right). In that test, the tip curved toward the light. From this, the Darwins concluded that when a seedling is exposed to light, some stimulus is transmitted from tip to stalk causing the latter to bend.

Some years ago, Professor Joseph F. Gennaro, Jr., and Terry Ashley of New York University raised fruit flies in covered dishes containing sugar solutions, yeast, and

Responses and Adaptations

radioactive phosphorus. This diet made the fruit flies radioactive.

The radioactive fruit flies were killed, then washed and placed on individual tentacles of a sundew plant. After several hours—days in some cases—the flies were removed. Then the plant was cut up and all parts of it were tested for radioactivity.

Sundew plants bear tentacles on the wide upper parts of their club-shaped leaves. The scientists were also able to place flies on leaves from which the tentacles had been clipped. These leaves were also later tested for radioactivity. The leaves from which the tentacles had been clipped did not absorb much of the flies' radioactivity. But the leaves that were intact had absorbed 80 percent of the flies' radioactivity within twenty-four hours.

Photographs of the plant with intact leaves showed that material from the radioactive flies was distributed first to the leaves and then through the plant to the growth center, or meristem. The growing cells of the plant took up most of the radioactive material initially. Only much later did the material appear elsewhere in the plant.

Dr. Gennaro and Mr. Ashley showed that the tentacles of the sundew plant are responsible for the digestion and absorption of trapped insects. They also showed that the sundew uses the products of this digestion for its growth processes.

In general, the respiratory and reproductive systems of

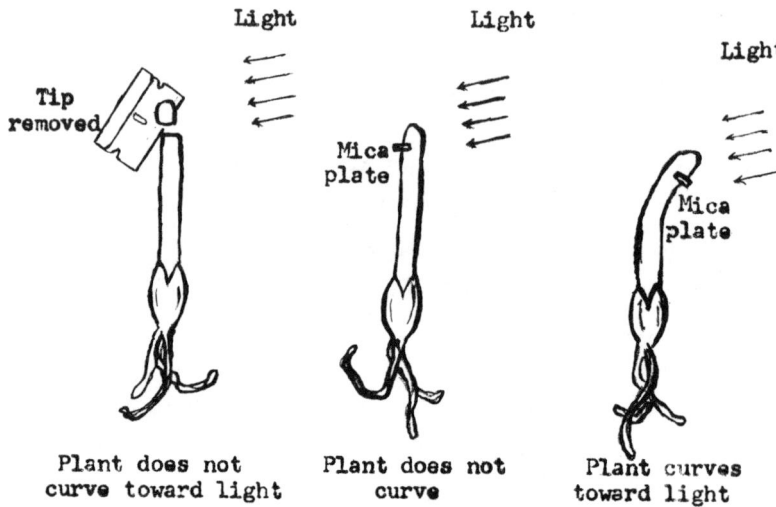

Plant does not curve toward light   Plant does not curve   Plant curves toward light

In 1910, Danish scientist P. Boysen-Jensen went a step further with the Darwins' experiment. Dr. Boysen-Jensen removed the tip of a seedling (left) and noted that it would no longer bend toward the light. Then he placed a sliver of mica in the shaded side of the stalk just below the tip (center). He found that the seedling would not curve toward the light. Then he placed a mica sliver on the illuminated side of the stalk (right), and he found that the seedling in this case curved toward the light. Dr. Boysen-Jensen concluded that the stimulus that prompted the plant curving toward light must travel down the shaded side of the stalk.

plants are very different from those of most animals. But there are certain areas of similarity. Sexual reproduction is a strategy used by all seed-bearing plant species, while asexual reproduction (budding) is not unknown among animals.

Plants need oxygen, just as animals do. But many

Responses and Adaptations

plants breathe out more oxygen than they breathe in.

So far, we've talked about plant responses and movements. We've discussed the similarities and differences of plants and animals. Yet we now seem to have more questions than answers.

It is all very well to say that plants move toward light and to call this response a tropism. It is fine to say that roots grow downward because they respond to gravity, or to say that long-day and short-day plants exhibit photoperiodism. But what does it all mean? What makes a plant grow toward the light? How does it "know" where the light is? Why does a plant respond to gravity or darkness by changing the way it grows? Must not a plant be in some way aware of its environment? If so, what is the nature of that awareness?

In thinking and talking about awareness, we tend to link it primarily to sight and sound images. We may also link it to such physical feelings as those of warmth and pain. The awareness of higher animals to their environment is localized in sense organs such as eyes and ears. Stimuli collected by these organs are transmitted to reception centers by nerves which trigger certain responses in the animal.

Animals do not depend directly on light for their food. Their primary attention is focused on other organisms which may provide them with food or threaten their safety. Animals may respond to temperature variations by building shelters, by hiding in thickets or crevices, by

PLANT MYSTERIES

hibernating, or by migrating to places with more suitable climates.

Fighting, seeking shelter, pursuing food, and fleeing from danger all require rapid responses. Therefore, animals have evolved a special system for receiving and processing information rapidly. That special system is the nervous system.

A few plant types, such as the carnivorous plants, are capable of rapid responses. But as far as we know, those plants or any other plants do not have nervous systems. Are nervous systems essential for awareness?

Some animal responses are below the threshold of sensory awareness. Normally, animals are not aware of the processes of digestion and metabolism. Animals are not conscious of the processes of cell growth and repair. They do not decide to mature sexually or to grow old. These processes are regulated within animal bodies by enzymes and hormones. Enzymes are catalysts—substances that speed up chemical processes that would occur slowly or not at all without them.

Hormones can be thought of as "chemical messengers." They "tell" the cells of various body organs what to do and how to grow. They stimulate or slow down the production and release of various substances. Though we are not consciously aware of what is happening as we absorb and use food, the cells of our bodies can be said to be "chemically aware" of the need to respond to the messages of the hormones.

Responses and Adaptations

Like the nervous system, the hormone system of animals is also a system that receives and processes information and causes responses to that information.

Since plants are sedentary (fixed location) creatures, they can afford to be more leisurely in their processing of information than animals. But plants can't afford to be insensitive to their environment. They need to monitor light and temperature conditions. They need to be aware of the seasons. They need to be aware of the availability of moisture. Plants need to know which way to grow. Some of them must be able to find support.

As a result, plants have developed a kind of chemical awareness of their environment. They contain various pigments in their bodies which respond chemically to light. They have developed a complex hormone system.

Biologists still don't fully understand how the complex hormone system of plants works. They realize that plants manufacture a number of substances that convey information and direct plant growth in ways that are analogous to the way in which animal hormones function. These substances make plant bodies chemically aware of their environmental conditions. These substances enable the plants to respond to the directions of chemical messengers.

Many of the exact mechanisms by which plants receive, transmit, and act on information are poorly understood at this time. That's not true, however, of the mechanism of phototropism. A series of experiments—

PLANT MYSTERIES

beginning with one by the Darwins—has given us a fairly clear picture of this plant response. And a more recent series of experiments has begun to show us how plants adjust themselves to seasonal changes.

CHAPTER 11
Light... and Gravity

OATS AND OTHER GRASSES have small hollow sheaths called coleoptiles. Each sheath protects the first leaf of the grass as it comes out of the ground. The coleoptile tip is extremely sensitive to light. It will bend toward a light source. Charles and Francis Darwin observed this, and they were fascinated by it.

The Darwins cut the tips off some grass seedlings. They found that remaining sections of the seedlings would no longer bend toward the light. The Darwins also covered the coleoptiles of some seedlings with opaque caps. That kept light from reaching the seedling tips and also prevented the seedlings from bending toward the light.

Something in the coleoptiles, the Darwins concluded, perceived light. That something relayed the light message to the cells just below each seedling tip. Then those cells grew longer on the dark side and curved the seedling blade toward the light.

PLANT MYSTERIES

As time passed, other scientists became interested in the subject. Those scientists found that whatever it is that "tells" the cells to grow can pass through gelatin. They also showed that this growth information can be collected and stored on a block of agar (a gelatinous substance).

A tip cut from a blade of grass was placed on a block of agar and exposed to light. Later, the tip was removed and the block of agar was placed on the beheaded blade of grass. Then the blade of grass started to grow again!

Scientists also found that when the agar block was placed on the side of the grass blade, instead of entirely across it, the side under the block grew faster. As a result, the blade of grass curved as it grew.

Obviously, some kind of chemical messenger was being produced in the grass coleoptile. This messenger was able to regulate cell growth. Scientists named the chemical messenger *auxin*.

In animals, substances known as hormones are produced in one body organ and act elsewhere. Since auxin was produced in one part of a plant and acted elsewhere, biologists considered auxin to be a plant hormone.

Light causes auxin to migrate from the tip of a grass blade to cells on the darker side of the blade and promote the growth of those cells. In that way, the tip or coleoptile of a grass seedling enables the seedling to curve toward the light. In a similar manner, root cap

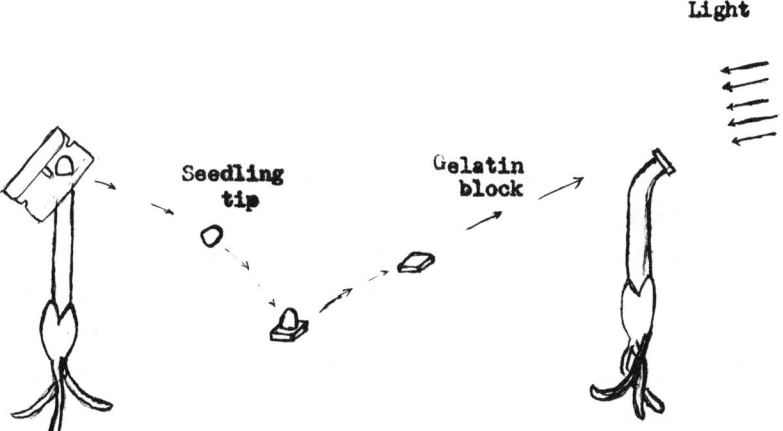

Dr. Frits Went, a Dutch botanist, elaborated on Boysen-Jensen's experiment in 1926. Dr. Went sliced the tips off several seedlings. He found that the "headless" seedlings stopped growing. Dr. Went then placed the cut-off seedling tips on small blocks of gelatin. He left them there for about two hours. Then he removed and threw away the seedling tips. He placed the treated gelatin blocks on tipless seedlings. The seedlings resumed growth. What's more, they curved toward a light source. Dr. Went also capped some tipless seedlings with plain gelatin. But those seedlings did not resume growth—they did not curve toward a light source. Dr. Went concluded that a chemical substance in plant tips responds to light and stimulates plant growth. At a later date, other scientists named the substance *auxin*.

cells in plant roots seem to be involved in causing a root tip to respond to gravity.

A group of scientists at Bedford College in London, England, gently removed the root caps of some corn

PLANT MYSTERIES

plants. The corn roots kept on growing, but they no longer curved downward in response to gravity. In fact, the roots did not curve downward until they had regenerated their root caps. In other words, the plants' roots had to grow new cap cells in order to regain their sense of direction.

Auxins control the response of a plant's root and stem to gravity. If a plant is placed on its side in the dark, the stem tip will soon turn upward. Its root tip will start to grow downward toward the center of the earth? Why?

In the stem, auxin moves to the lower side of the stem. It causes the cells there to elongate (grow long). That cell growth causes the stem to curve upward. Auxin also moves to the lower side of the root. But its action there is different. Instead of stimulating the cells on the lower side of the root to elongate, it stops them from doing so. Then the cells on the upper side of the root elongate because the auxin doesn't inhibit them. As a result, the root curves downward.

Professor Arthur W. Galston believes that phototropism (the bending response to light) and geotropism (the bending response to gravity) are both associated with electrical activity in plants. Many years ago, Professor E. J. Lund of the University of Texas showed that there are fewer electrons near the tip of a grass seedling than at its base.

If a leaf sheath, stem, or root is laid on its side, it develops a very small electrical charge that can be meas-

Light ... and Gravity

ured. The same thing happens if the plant part is strongly illuminated from one end. In the leaf sheath, the electrically positive side starts to grow faster than the rest of the sheath. That side grows by cell elongation. This unequal rate of cell elongation is, as we have seen, the mechanism of phototropism and geotropism.

A. R. Schrank, another scientist working at the University of Texas, found that he could prevent these electrical charges from developing. He could stop plants from curving in response to light or gravity by passing a direct electric current through the plants. But this response depended on the direction of the current flow. By reversing the positive and negative electrodes, Mr. Schrank found that he could increase a plant's curving response to light or gravity.

The work of Lund and Schrank made it clear that electrical activity plays some part in plant tropisms. But other research had established fairly strongly that the movement of auxins is involved in phototropism and geotropism.

Could the two be linked? Could the movement of auxins and electrical activity be interrelated in plant growth movements?

Two Swedish scientists, Lennart Grahm and E. H. Hertz, showed that auxin must be present for the electrical voltages that accompany the curving responses of plants to develop. The movement of auxin comes before the appearance of the electric potential. The scientists

PLANT MYSTERIES

speculated that auxin may cause the build-up of electric potential by speeding the transfer of charged particles across cell membranes.

In any case, it's obvious that auxin is the chemical messenger. But what picks up the message to send the messenger?

George Wald, the discoverer of vitamin A in the retina of the eye, believes that carotene is the light sensor for phototropism.

In a *Scientific American* article called "Life and Light," Dr. Wald wrote that plants primarily use long light waves (red light) for photosynthesis. But these long light waves have nothing to do with phototropism. Measurements with monochromatic (one-color) lights showed that blue-green, blue, and violet light waves cause phototropism. As you know, these waves are near the short end of the visible light range.

Dr. Wald concluded that while green plant pigments absorb long-wave light for the process of photosynthesis, yellow plant pigments absorb short-wave light for phototropism.

All phototropic plants, Dr. Wald reported, seem to have yellow pigments. These pigments are called the carotenoids. In some plants, these yellow pigments are located in the precise parts of the plants that show phototropic responses.

But plants are not the only organisms that possess carotenoids.

As far as a plant is concerned, all light is not alike. Two similar plants were exposed to two different artificial light sources in a Westinghouse laboratory to gain knowledge of the light requirements of indoor plants. Both plants are three months old. However, the one on the left was illuminated by a specially developed Agro-Lite fluorescent lamp. The plant on the right was illuminated by a more conventional light source. (*Westinghouse*)

PLANT MYSTERIES

Many sea animals, such as hydroids, are attached to the ocean floor by stalks. These animals resemble plants, and like plants, many of these sea animals grow toward the light. Phototropic responses in these animals seem to be stimulated by the same short light waves that stimulate phototropism in plants. Dr. Wald believes that carotenoids in these sea animals are responsible for their phototropic responses.

In a more remote way, carotenoids seem to be linked to the growth, vision, and health of higher animals. Vision in higher animals is dependent on an adequate supply of vitamin A. That's because the pigment, visual purple, which is necessary for sight, is a compound of vitamin A and a protein.

Humans—and other higher animals—who don't get enough vitamin A are said to suffer from "night blindness." They have particularly poor vision in dim light.

Animals cannot make vitamin A in their bodies. They get it from plant carotenoids that they eat. Therefore, since carotenoids are responsible for plant phototropism, there appears to be a connection between plant phototropism and animal vision.

Though much remains to be learned, scientists now know many details of the sensing mechanism for phototropism in plants. Much less is known, however, about the sensing mechanism for geotropism.

Several scientists think that small particles in plant cells are responsible for the sensing of gravity. These

Light... and Gravity

particles are tiny, mobile grains of starch. They are called statoliths. Within a plant cell, such a particle may tumble from one cell wall to the other when the cell is tipped. In this way, the cell can sense which way is "up."

Some animals use a similar mechanism to sense gravity. But, in animals, the statoliths are enclosed in spheres lined with sensory hairs. In humans, statoliths are contained in a jellylike mass located in special chambers within the inner ear. There, statoliths help people to maintain their balance.

Scientists are now trying to find out precisely how statoliths control geotropism in plants. They want to know what, if anything, statoliths have to do with the movement of auxin in plant tissues and the development of electric potentials in plant bodies.

Though scientists are learning much about the mechanisms of phototropism and geotropism, the other tropisms are still shrouded in mystery.

Some tendrils, the slender clinging organs of certain vines, show not only thigmotropism, but negative phototropism as well. These tendrils respond to contact with a solid object by forming little adhesive disks that fasten the vine to the object. This response is a thigmotropism. But another response seems to be involved in this. The disk-making vines actually seek out objects to support them. They seem to do this growing away from light toward darker areas. This response may be negative phototropism or it may be skototropism.

PLANT MYSTERIES

In either case, scientists don't know what sensing mechanisms are involved. They wonder what role auxin may play in this.

Plants can also direct their growth in response to water, heat, and certain chemical substances. A pollen tube, which starts out on the stigma of a flower from a pollen grain, grows downward through the style. It follows a route marked for it by certain chemical substances in the style. So, the pollen tube can be considered an independent organism showing a chemotropic response.

This leads us to some important considerations of the responses and nature of plants. If plants "know" where light is and can turn themselves toward it, if they can change their direction of growth in response to gravity, if they can follow chemical paths, then they are not insensitive to their environment.

Plant tropisms, which involve bioelectric activity and the chemical messengers called hormones, show the existence of a kind of plant information system. This system contains sensing devices—pigments and statoliths. It contains transmitters—plant hormones and related electric activity. It has "effectors" of the information. These are the actual cells that respond to the presence of hormones by elongating—or by not growing at all.

Because a definite response to an environmental stimulus is produced by the chemical and bioelectric transfer of information in this system, we are tempted to compare this system to an animal nervous system. After all,

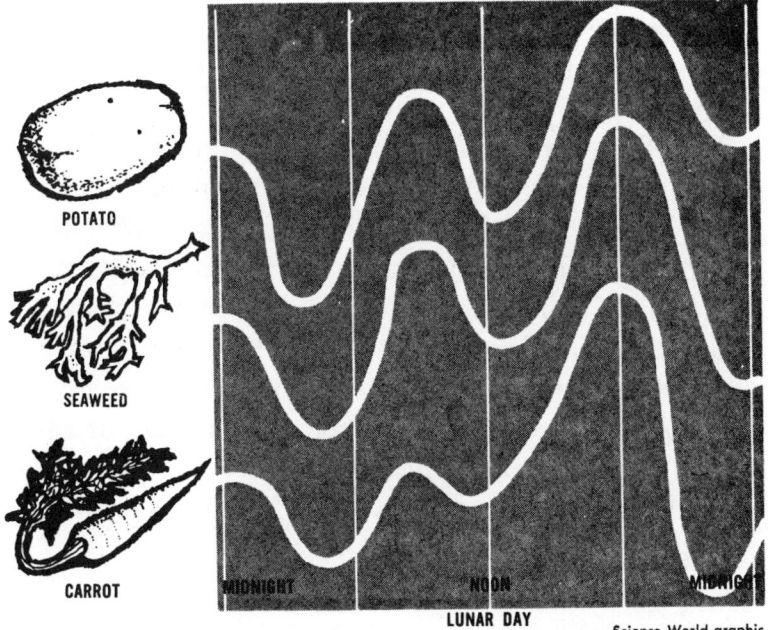

Different plants show a similarity of metabolic functioning during a "lunar day"—a period determined by the motion of the moon. The moon is highest in the sky at noon, lowest at midnight. This and similar observations suggest that biological rhythms are influenced by changes in gravity, magnetism, and perhaps air pressure.

both systems are communication systems within living organisms. But we must remember the differences between the two systems. Plant tropic responses are usually much slower than animal nervous responses. Plants respond to a much more limited range of stimuli than do animals.

PLANT MYSTERIES

Plants sense light. But they have no way of forming images as animals do with their eyes. There is no evidence that plants feel stimuli.

The effectors in an animal nervous system are in central regions. They are specialized for responding to messages. Higher animals have a central organ called the brain which makes decisions and can send messages and directions to other parts of the body. But the primary response in the effectors (the cells) of a plant information system seem quite limited. For instance, a cell can grow, or stop growing, in response to a message. It can release an enzyme. Or it may empty itself of water. But it does not appear to have much decision-making capacity.

The responses of plant cells seem to be like the growth responses of animal cells. Those animal cells are also controlled by messenger hormones. The responses of plant cells can't really be compared to animals' nervous responses.

Scientists have found no way in which a plant could sense a thought or an emotion directed toward it by another living thing.

Plants are chemically aware of their environment. But they are not conscious!

CHAPTER 12
A Chemical Awareness

PLANTS' CHEMICAL AWARENESS is much more complex than scientists had once thought it to be. This awareness is not limited to growth movement responses—the well-known tropisms. Plants are chemically aware of the seasons. They appear to have their own biological clocks.

Many plants regulate their growth and reproduction according to the seasons of the year. In a general way, people have long been aware of this. For example, people saw that tobacco plants usually flower in the summertime. They assumed that cold winters and spring planting times accounted for this. However, two scientists in the 1920s found that more was involved than that. The scientists discovered photoperiodism in plants.

W. W. Garner and H. H. Allard found a mutant strain of tobacco that would flower only in December. In other words, they found a short-day tobacco plant even though most other tobacco plants are long-day plants. The two scientists thought that they had dis-

PLANT MYSTERIES

covered a plant response based on the length of the light period, or day. They called the response photoperiodism.

Some years later, two other scientists started to study photoperiodism using the cocklebur, a very tough plant. Pluck off all its leaves, and it will still live. The cocklebur is also a short-day plant. One short-day cycle—exposure to a long night—is enough to make the cocklebur flower. That will happen even if the plant is then returned to long-day cycles.

The scientists, Karl C. Hammer and James Bonner, soon discovered that if the cocklebur's long night of darkness was broken by even a brief flash of light, it would not flower. But interruption of the plant's short day by a period of darkness would not prevent flowering.

From this, the scientists concluded that short-day plants are not measuring light—they are measuring darkness!

The scientists found that long-day plants also measure darkness. A single flash of light interrupting a long night was enough to make a long-day plant flower.

Hammer and Bonner also showed that it is the leaves of a plant that respond to light. When all the leaves were stripped off a cocklebur, the plant would not flower. But when a small piece of a single leaf was left on another cocklebur, the plant flowered.

Scientists found that by flashing red light of wavelength 660 nanometers (660 billionths of a meter) at a cocklebur, they could most effectively prevent the plant

A Chemical Awareness

from flowering. They also found that such red light also prevented the flowering of other short-day plants.

However, the same wavelength of red light, 660 nanometers, was found to be most effective in stimulating the flowering of long-day plants. This light was also found to be effective in stimulating lettuce seeds to sprout.

Then it was discovered that a flash of far-red light—730 nanometers in wavelength—could stop the germination (sprouting) of lettuce seeds. Furthermore, flashes of far-red light were found to reverse the effects of red light—660 nanometers—on short-day and long-day plants.

The scientists soon realized that they could turn on and off the flowering responses of plants simply by flashing red or far-red light at them. Why? Some scientists suspected that a pigment, or pigments, was involved in these responses. The search for such a pigment was on.

Finally, S. B. Hendricks and H. A. Borthwick of the U.S. Department of Agriculture identified the responsible substance—phytochrome.

Phytochrome consists of a blue pigment, phycobilin, and a protein. It exists in two forms that are usually called P_{660} and P_{730}.

P_{660} is a more stable substance than P_{730}. It absorbs red light from the sun and it is changed by that light to P_{730}.

P_{730}, which absorbs far-red light, prevents flowering in short-day plants, and promotes flowering in long-day plants. During periods of darkness, P_{730} gradually changes back into P_{660}.

Scientists concluded that relative amounts of the two forms of phytochrome in a plant trigger or inhibit the plant's growth. But the scientists are not sure how exactly this happens. In fact, plant biologists disagree on many details of this mechanism.

One thing is sure: the phytochrome mechanism triggers different responses in different plant species. It seems that in each species, the phytochrome mechanism causes the response which is best adapted to ensure the survival of that plant species.

Phytochrome acts as a light-sensing substance in plants. It plays a role in the opening and closing of flowers and it helps bring on dormancy in plants. It seems to be involved in the dropping of leaves in the fall.

However, phytochrome does not work alone in triggering these responses. Scientists believe that several plant hormones, and perhaps some enzymes, may be involved as well. Some scientists think that phytochrome changes cell membranes so that hormones can pass through them more or less easily. But there are several other theories.

The role that phytochrome seems to play in leaf movement is a good example of how complicated the study of plant information systems can get. In a series of articles in *Natural History* magazine, Professor Arthur W. Galston explained what seems to happen.

A Chemical Awareness

The leaves of many plants seem to go to sleep at night. They may fold themselves against opposite leaves or against the stem of the plant. Charles Darwin thought that plants might do this to cut down on heat loss by showing as small a surface area as possible. By presenting such a small area, the leaf would lose relatively little of the heat it had absorbed during the day.

Professor Galston reported a different opinion. He wrote that a German scientist, Erwin Bunning, now thinks that photoperiodic plants may fold their leaves to escape the moonlight. Such plants are very sensitive to light. Even bright moonlight, Bunning suggested, might stop short-day plants from flowering and cause long-day plants to flower too soon. By folding their leaves, these plants offer as small a light-absorbing surface as possible.

Several other scientists are not sure about Bunning's theory. Professor Galston thinks that Bunning and Darwin may both be partly right.

Many "sleeping" plants open as soon as light shines on them. Many "awake" leaves fold up as soon as they are moved into darkness. Since phytochrome operates as a light-sensing pigment to control flowering, we might expect it to also affect light-related leaf movements. In fact, phytochrome does help control these movements, but only when the plant's biological clock is set in the right position.

If a plant which normally folds its leaves in response to darkness is left in the light long enough, it will fold

PLANT MYSTERIES

its leaves anyway. If such a plant is kept permanently in the light, it will open and fold its leaves at twelve-hour intervals. A plant continuously kept in darkness will also open and close its leaves at regular intervals. The whole cycle will take about one day.

Obviously, phytochrome has nothing to do with such a cycle. But what is involved?

Some interior "clock" seems to help the plant establish a daily rhythm. This clock is independent of environmental conditions. Scientists don't know how it works.

The closing of leaves in response to the daily interior rhythm seems to be a different, more passive process than the closing triggered by darkness. This rhythmic closing can occur in the presence of substances which prevent the dark-triggered closing.

Phytochrome does regulate leaf closing when the daily rhythm of the plant allows such closing to occur. Plants exposed to a period of natural light normally respond to darkness by closing their leaves. But such a plant will not fold its leaves in the normal way if it is exposed to far-red light before being placed in darkness. Why?

The accumulated P_{730} pigment in the plant's tissues is changed back to P_{660} pigment by the exposure to far-red light. Treatment of the same plant with normal red light enables the plant to close its leaves rapidly in darkness.

During such leaf movements, electric potentials are

A Chemical Awareness

produced in the plant tissues. Richard Racusen of the University of Vermont and Ruth Satter of Yale placed electrodes in the plant cells responsible for sleep movements in leaves. The scientists found differences in electric potentials between the different parts of the cells which varied according to the time of day and the light conditions outside.

Several scientists concluded that the sleep movements of leaves are controlled by the movement of potassium in and out of cells in the fleshy joint which joins the leaf stalk to the stem. When the upper cells contain much potassium, they absorb water and swell up. This pushes the leaf down and out to the open position. As potassium moves to the lower cells, they swell up and push the leaves closed.

How does phytochrome regulate this process?

M. J. Jaffe of Ohio University suggested that phytochrome may somehow control the release of a substance called acetylecholine. If this substance is released in the vicinity of the control cells, it may cause potassium movement.

This theory is fascinating because acetylecholine is the substance that transmits messages from one cell to another in animal nervous systems. What's more, animal nervous systems rely on sodium and potassium exchanges to transmit information from one nerve cell to another.

So, if this theory is correct, it suggests a certain basic similarity between plant information systems and animal nervous systems. But is this theory correct? Some findings support it. Others do not.

Acetylecholine has been definitely identified in plants. Other substances that occur in animal nervous systems have also been found in plant systems. Several close chemical relatives of substances in animal systems exist in plant systems.

Auxin, a very important plant hormone, is chemically related to serotin, one of the substances that helps transmit nerve impulses in animals.

In animals, no absolute line can be drawn between the hormone message system and the nerve message system. Nerve cells in animals secrete certain hormones. They also stimulate the secretion of other hormones.

We should not be surprised if it turns out that a plant hormone is found to resemble the substance that transmits a nerve impulse in an animal. Neither should we be surprised if it is found that the electric potentials linked to the sleep movements of leaves are based on the movement of potassium ions (charged atoms). The movement of potassium ions is used by animal nerve cells to transmit messages.

Partial evidence indicates that plant cells may use some of the mechanisms used by animal nervous systems to transmit messages. Or, plant cells may use similar mechanisms. Scientists hope to gain a better understanding of this by studying plant hormones and the role of these hormones in regulating plant growth.

CHAPTER 13
Plant Hormones

AUXIN IS A PLANT HORMONE produced by the growing tips of stems and roots. From these tips, auxins travel back into the stems and roots where they cause or inhibit cell elongation. But that's not all. Auxin plays a role in several other plant responses.

Auxin produced in the primary (central) bud of a plant causes that bud to grow rapidly. But it inhibits the growth of the lateral (side) buds.

This tendency of the central bud to grow more rapidly is known as apical dominance. Pinching out the central bud will, however, permit the side branches to grow. Gardeners often pinch off central buds and flowers to obtain bushier plants. They prune bushes for the same reason.

In experiments, scientists have cut apical (central) buds off plants. Then they have applied auxin to the remaining cut end of the branch. They've found that the side buds react as if the central buds were still there. The side buds do not grow.

PLANT MYSTERIES

Auxin also prevents plant leaves and fruits from dropping too soon. A healthy leaf or developing fruit produces auxin. The hormone acts to keep the leaf or fruit firmly attached to the plant stem. However, as the leaf or fruit ages, it produces less and less auxin. As that happens, a special layer of cells begins to form where the leaf, or fruit, connects to the stem. This is called the abscission layer.

The purpose of the abscission layer is to weaken gradually the connection between the leaf, or fruit, and the stem. Eventually, the abscission layer becomes dry and brittle. The slightest movement will then shake or jar the leaf or fruit loose.

Growers who want to keep leaves or fruit on a tree can spray the tree with a synthetic (manmade) auxin. This stops the development of the abscission layer. The fruit usually remains on the tree until the farmer is ready to harvest it.

Auxin can help growers in other ways. It can be used to bring on flowering in pineapple plants. It can be applied to the unpollinated blossoms of some fruit trees to produce fruit that are normal in every way but one—they contain no viable seeds.

Because growers can use auxin for such purposes, scientists figure that auxin must somehow influence flowering and fruit setting in plants. Pollination must, for example, somehow cause an increase in the auxin present in the ovary of the blossom. This increased auxin then

Plant Hormones

probably causes the setting and development of the fruit.

Auxin seems to foster cell division. It causes the emergence of branch roots from the main root. In this action, the auxin seems to work with another group of substances—the cytokinins.

Cytokinins occur in seedlings, fruitlets, and in coconut milk and other liquid endosperms (nutritive tissue in embryo sacs). The function of these substances has been examined by several scientists, including F. C. Steward.

Dr. Steward performed a series of experiments in the culture of plant tissue. He placed mature cells from a nongrowing portion of a carrot root in a growth medium. Then he added coconut milk to them. After twenty days, Dr. Steward found that the mass of carrot cells had multiplied eighty times.

First, the cells increased in size. Then they increased in number. As the cells increased in number, the size of the individual cells decreased. When the process of cell division slowed down, the size of the individual cells started to increase again.

It is important for us to remember that these cells would never have grown again in their natural state in the carrot root. Obviously, the coconut milk was promoting the growth of the cells in their unnatural environment.

As Dr. Steward continued his experiments, he found that synthetic auxin helped the coconut milk to cause the carrot cells to grow and divide. Later, he found sub-

PLANT MYSTERIES

stances other than coconut milk that would stimulate cells to grow. These substances include extracts of immature corn grains, walnuts, and horsechestnuts.

Since all these substances contain cytokinins, this seemed to establish the role of kinins in the growth and division of plant cells.

Scientists found that the synthetic substance kinetin, which is closely related to the natural kinins, also promotes the growth of plant cells. When this substance is applied to cut leaves, it keeps them green. If it is placed on a small area of a cut leaf, that area will stay green while the rest of the leaf turns yellow.

Kinetin seems to cause plant nutrients to build up in the area where it has been applied. Because of this ability, some scientists consider kinetin and the kinins to be plant hormones. But Professor Arthur Galston disagrees. Since they act locally, he is unwilling to consider the natural kinins or kinetin as hormones.

Dr. Steward investigated another class of plant-growth enhancers—the gibberellins. These substances were first identified in a fungus that infected rice plants. Scientists noted that something in the fungus caused the plants to elongate and die prematurely. Researchers isolated the responsible substance, and they named it gibberellin. Later, they found that dwarf trees treated with the substance would grow tall. They also found that long-day plants treated with it flowered.

Gibberellin can stimulate seed germination. It can

The effects of various hormones on plant growth were studied by Dr. John W. Mitchell of the U. S. Department of Agriculture in the late 1960s and early 1970s. This photo shows some results of hormone treatment. Bean plant on left was untreated. Plant in center was treated with gibberellin. Plant on right was treated with brassins. (*USDA*)

spur the growth of grasses. It can shorten or replace the cold period that some plants need before they can germinate or resume growth.

Scientists don't know for sure how this growth enhancer works. They suspect that the gibberellins somehow start the production of certain enzymes in plants.

Gibberellin probably acts with the cytokinins to end the dormancy, or rest period, of seeds and buds. Phytochrome may be involved in altering the levels of gibberellins and cytokinins in seeds and buds. Temperature changes may also help alter the levels of these substances.

As we've seen, plants contain growth enhancers. But they also contain certain growth preventers. This can be demonstrated with a simple experiment. Some seeds can be made to germinate by placing them under running water for several hours. This imitates the action of a heavy rainfall in nature. It works by washing or leaching growth-regulating substances out of the seeds. When these substances are thus removed, the seeds will germinate if there is enough water present to foster their growth.

One growth regulator that might be washed out this way is abscissic acid. It is a plant hormone that causes the formation of the abscission layer of cells at the separation point between leaves, or fruit, and plant stems. As we saw earlier, this abscission layer promotes the dropping of a leaf or fruit.

Abscissic acid also helps bring on dormancy in seeds

Plant Hormones

and buds. The amount of abscissic acid in the seed or bud often decreases as the length of exposure of the plant to cold increases. At the same time, gibberellins and cytokinins build up in the seed or bud. This mechanism probably enables plants to be "aware" of and respond to temperature changes.

These hormones and related substances seem to work together to control plant growth and to produce responses to environmental changes. Scientists don't fully understand how all these substances work together. There are gaps in our understanding of the system. These gaps are probably hormones that have not yet been identified.

When these hormones are identified, scientists may have a clearer picture of plant sensitivity to environmental conditions. They may also be able to control and alter various plant responses to these conditions.

So far, a number of artificial plant hormones have been produced in laboratories. Several more will probably follow. Then scientists will more and more be able to artificially regulate the flowering, ripening, and growing habits of agricultural plants. They may also be able to equip these plants to withstand environmental conditions that might otherwise damage them.

CHAPTER 14
Plants and Sound

PLANTS ARE DEFINITELY SENSITIVE to light, gravity, and temperature change. Some scientists think that plants may also be sensitive to sound. But that is a more controversial idea.

In *The Secret Life of Plants*, Peter Tompkins and Christopher Bird describe the experiments of T. C. Singh, a botanist at Annamalai University in India. Dr. Singh's first experiment, the authors report, was suggested by the well-known British biologist Sir Julian Huxley.

Huxley saw Dr. Singh observing, through a microscope, the streaming of protoplasm in an Asian water plant called hydrilla. Sir Julian wondered if the streaming of the protoplasm would be changed by sound. He mentioned it to Dr. Singh, and they decided to try to find out.

Both scientists knew that protoplasm streaming in plants usually increases after sunrise. Dr. Singh decided

PLANT MYSTERIES

to see if sound would make the protoplasm stream more quickly than usual. He set up a tuning fork six feet away from the hydrilla. He sounded the fork for a half hour prior to 6 A.M. He noted that the plant's protoplasm streamed much more rapidly than was normal for that time of day.

Dr. Singh also found that the hydrilla appeared to respond to violin music of a certain pitch.

Would other plants respond to music?

Dr. Singh started playing ragas—traditional Indian music—to various plants in a controlled experimental setting. Ragas were played to a large number of plants of different species for half an hour every day just before sunrise. The music was generally high in pitch. Instruments used for playing the ragas included a violin, a flute, and an Indian stringed instrument called a veena.

Plants exposed to this music, Dr. Singh found, grew better than the control plants that were not exposed to music.

Following that experiment, Dr. Singh played ragas to six varieties of rice growing in fields. He got yields of rice from those fields that were from 25 to 60 percent greater than yields normal for that region.

As far as he was concerned, Dr. Singh was convinced that he had shown that sound waves could affect the growth of plants. He offered no mystical or psychic explanation for the apparent effect of music on plant growth. He considered that transpiration and carbon

Plants and Sound

assimilation—two fundamental parts of plant metabolism—are speeded up by musical sound and rhythmic beat. The music, Dr. Singh reasoned, stimulates the plants to make more food for themselves, and thus grow faster.

Of the people who heard of Singh's experiments, several were impressed. That included George E. Smith, an American researcher working for a wholesale seed supplier. Mr. Smith decided to try some experiments on the effects of sound on plants. According to Bird and Tompkins, Mr. Smith planted corn and soybeans in each of two greenhouses. Conditions, particularly temperature and moisture, were kept as similar as possible in both greenhouses.

Then Mr. Smith set up a record player in one of the greenhouses. For twenty-four hours a day, the corn and soybeans there were exposed to the melody of George Gershwin's "Rhapsody in Blue." Meanwhile, quiet prevailed in the other greenhouse. It was the control.

Periodically, Smith checked the plants' progress in both greenhouses. He found that the plants exposed to music germinated sooner than the control plants. At various times, he also reported, plants exposed to the music weighed more than those from the quiet greenhouse.

This would seem to indicate that music—at least Gershwin's "Rhapsody in Blue"—spurs the growth of corn and soybean plants.

Mr. Smith also claimed that he could improve the

PLANT MYSTERIES

yield of hybrid corn in fields by broadcasting music to the plants.

Actually, exposure to continuous high and low notes, Mr. Smith reported, had more effect than melodies in increasing crop yield. The increases seemed to be due to better survival rates among the plants rather than to increases in the yield of each plant.

Like Dr. Singh, Mr. Smith offered no mystical or psychic explanations for these apparent effects of sound on plants. He suggested that the sound waves might increase molecular activity in corn.

Another researcher had heard of Dr. Singh's experiments, but she was not impressed. Canadian biologist Dr. Pearl Weinberger did not think that Dr. Singh's experiments were scientifically valid. She considered that the laboratory controls were not set up in a scientific fashion, and the results were not analyzed according to statistical methods.

Dr. Weinberger decided to run her own experiment. She chose Rideau wheat for her tests.

Dr. Weinberger placed chilled, water-soaked seeds in special chambers with the best possible conditions for sprouting. She broadcast sound waves of 5,000 hertz (cycles per second) at one group of seeds. She broadcast sound waves of 12,000 hertz at the other group of seeds. The third group of seeds, the control group, received no sound treatment. At intervals, Dr. Weinberger measured and weighed the seedlings. She counted their

Plants and Sound

roots and leaves. She repeated this experiment ten times.

The seedlings exposed to sound waves of 5,000 hertz grew to be 250 to 300 percent heavier than the control seedlings. The seedlings exposed to 12,000-hertz sound were 20 to 50 percent heavier than the control seedlings.

Dr. Weinberger concluded that sound waves may somehow affect metabolism in plant cells.

George Milstein, a retired dental surgeon, also experimented with plants and sound. He found that a continuous hum at 3000 hertz could make tropical plants grow better than normal and bloom ahead of schedule. In fact, Dr. Milstein and a recording company produced and marketed a record of sound vibrations to stimulate plant growth. The sonic hum was masked by music and the record was labeled "Music to Grow Plants By."

The public response to the record and to his experiments baffled Dr. Milstein. He was deluged with letters and phone calls. Many of them came from people who considered that Dr. Milstein had proved that plants enjoy music. That notion shocked Dr. Milstein. He had meant no such thing. He did not conclude that plants can hear or that plants have emotions. He did not think Cleve Backster's work was scientifically valid.

Several other experimenters exposed plants and seeds to sound. *Science News* magazine reported that four scientists at the University of North Carolina at Greensboro found that plants are affected by noise.

The scientists involved were physicists Gaylord T.

PLANT MYSTERIES

Hageseth and Gerald W. Meisner, chemist Irma L. Morrison, and biologist Ralph M. Morrison. They exposed dry and wet turnip seeds to "pink" noise, a mixture of sound waves of different frequencies, but perhaps not so varied and jumbled as "white" noise. The sound waves of this "pink" noise ranged from 20 to 20,000 hertz in frequency. The sound level ranged from 100 to 110 decibels—quite a racket! In addition to the seeds that were exposed to this very loud noise, control seeds were kept in a similar environment, but the sound level there was no greater than sixty decibels.

The scientists found that the dry seeds were not affected by the experimental racket. But the wet seeds exposed to the loud noise germinated about 10 percent more rapidly than the control wet seeds. The wet seeds exposed to the loud noise had a germination rate that was 100 percent greater than that of the controls. In other words, the very loud noise seemed to cause practically all of the wet seeds to germinate. Why?

The experimenters were not sure. But they offered two theories. Somehow, the loud sound waves may break down the hard coats of the turnip seeds so that oxygen and water can enter the seeds and trigger sprouting. However, that was not the favored theory of the researchers. They thought it more likely that energy is transferred from the sound waves to standing waves in the seed cells. By increasing energy in the cells, the sound waves would thus stimulate growth.

Do plants have feelings? Can plants enjoy music? Following publication of the book, *The Secret Life of Plants*, and the record, *Music To Grow Plants By*, countless science fair entries included "experiments" similar to the one shown. The girls played soothing music to the plant on the left and yelled abusive language at the one on the right. (*The Hartford Courant*)

PLANT MYSTERIES

Other noise experiments were performed with growing plants. A Drexel University physicist, Dr. Arthur Lord, and several of his students grew a dozen coleus plants in an environmental chamber. After measuring leaf length for eleven days to calculate the normal growth rate of the coleus, they exposed the plants to noise of 100 decibels. They found that the plants' growth rates decreased an average of 47 percent in six days.

Professor Lord concluded that the noise increased the rate at which plants give off water and decreased the amount of carbon dioxide available for photosynthesis.

Bird and Tompkins stated that three other North Carolina researchers had decreased the growth rates of twelve sterile male tobacco plants 40 percent by bombarding them with random noise.

Probably the most publicized research on plant responses to sound has been done by a former organist and singer, Dorothy Retallack. Her experiments were reported in considerable detail in *The Secret Life of Plants*.

While studying for a degree in music, Ms. Retallack took a course in biology that required that she prepare and run an experiment. Ms. Retallack had read about George Smith's experiments. She resolved to find out what sound might do to plants.

After running some preliminary experiments with another student, Dorothy Retallack proposed a formal experiment to her biology professor, Francis E. Broman.

Plants and Sound

He agreed. She was given the use of three environmental chambers.

One chamber was used to house a control group of plants. The other two chambers were used to hold experimental groups. Then, an F note sound was played steadily for eight hours to plants in one of the experimental groups. The same note was sounded intermittently for three-hour periods to the plants in the other group.

The plants exposed to the F note sound for eight hours a day died in two weeks. But the plants exposed to just three hours of the sound outgrew the plants in the control group.

What, some people wondered, was the significance of those findings? Ms. Retallack's experiment became a source of controversy at her school.

Two other students decided to conduct a more significant experiment with plants and music. They put groups of plants in two environmental chambers. They played classical music to one group. They played hard rock to the other.

The plants exposed to classical music were reported to have grown toward the radio. Those exposed to hard rock grew away from the radio.

Ms. Retallack then ran more experiments. All of them indicated that plants were harmed by exposure to rock music. Plants exposed to rock music apparently used

PLANT MYSTERIES

more water than those exposed to classical music. Yet, the roots of the plants subjected to rock music grew more slowly than the roots of control plants.

Ms. Retallack experimented with other kinds of music. She found that the musical instruments used seemed to make a difference. Her plants leaned toward stringed instruments but away from percussion instruments. Plants responded favorably to the sounds of an Indian stringed instrument called a sitar. They also leaned toward a phonograph speaker from which jazz music was coming. Ms. Retallack found that different musical sounds appeared to affect the evaporation rate of water in the environmental chambers.

By 1970, Ms. Retallack's experiments were receiving nationwide attention. She and Professor Broman prepared a report of the experiments and sent it to *BioScience*—the journal of the American Institute of Biological Sciences.

BioScience did not accept the paper!

The sounds produced in the performance of a musical score are complex and varied. It is almost impossible to analyze precisely the effects of such varied sounds. No one could be sure which sound waves caused an apparent change in plant growth. Such experiments could easily lead people to assume that plants hear music and respond to it emotionally.

In the reports on her early work, Ms. Retallack does

not give the impression that she thought that plants respond emotionally to sound. She speculated that the percussive sounds of rock music might somehow harm plants. She suggested that the sound of different wavelengths might somehow affect plants' use of water.

However, as her experiments went on, Ms. Retallack seemed to believe that plants "care" about the sort of music played to them. This notion is evident in her book, *The Sound of Music and Plants*. The book also discusses mystical feelings of plants and plant spirits.

Did the rejection of Ms. Retallack's early ideas by the scientific community turn her to such mystical ideas? Or did her experiments, such as the one in which she exposed plants to the music of the sitar, lead her to those notions?

Whatever the answer, Ms. Retallack's interest in ideas of a mystical nature of plants caused scientists to doubt the value of her experiments. Most of them considered that those experiments provided no new information about the responses of plants to sound. Many scientists are still not sure that plants can respond to sound at all.

Some reputable scientists, such as Dr. Weinberger, have exposed plants to sound. They have observed growth changes which, they think, were caused by the sound. But they cannot really explain how sound might affect plants. The best they can do is offer suggestions.

Plant physiologists have not identified a structure in

PLANT MYSTERIES

plants that seems to be sensitive to sound. Some scientists suggest that sound may alter plants' metabolism, but again, they have not been able to explain *how*.

Except for Ms. Retallack's findings, no other experiments have shown tropic (growth movement) responses to sound.

Scientists are not sure that sound doesn't affect plants in some way. But they haven't proved that it does.

Biologists are now fairly certain that humans and animals are affected by sound, ultrasound, and infrasound in many complex ways. Our health and our sense of well-being can be affected by sounds we hear and even by sounds we can't hear. Since all life is, in a sense, related, it is reasonable to suggest that if animals are affected by sound waves, plants are also likely to be affected by those waves.

CHAPTER 15
A Question of Sensitivity

ARE PLANTS REALLY VERY SENSITIVE?

The answer to that question depends on what we mean by "sensitive." If we mean that plants are able to perceive certain stimuli and react to those stimuli, the answer is yes!

Plants sense light, gravity, contact, and chemical stimuli. They react by moving their bodies toward or away from those stimuli. They respond differently to different kinds of light.

Plants also respond to temperature changes. They time their growth and flowering in accord with the changes they sense. As we have seen, they do all this by chemical means.

The chemical means used by plants for responses resemble the chemical communications systems of animals, including those of human beings.

Individual plants are sensitive to their environment.

PLANT MYSTERIES

So also, in a special way, are plant species and families of species. Collectively, plants of a species adapt to certain environmental conditions, and they respond to long-term changes in those conditions.

Animal species also have this collective sensitivity. They adapt to environmental conditions, and they respond to changes in those conditions. We call this process "evolution."

The mechanisms of evolution are very complicated. Scientists still argue hotly over details of just how evolution works.

In general, evolution is a process of selection. Those plants and animals that find the most successful living strategies for their environment will survive long enough to breed. By breeding, they pass on these strategies to their descendants. In other words, plants and animals that develop ways to cope with a particular environment get to live, reproduce, and pass on the "knowledge" to their offspring.

Plants and animals have adapted to varied environments in very diverse ways. For instance, the spines and thick water-holding leaves of cacti are a response of the species to the dry, hostile environment of the desert. Some plants species repel insect predators with repulsive odors. Other plant species use fragrant scents or nectar to lure insects to pollinate their flowers.

These and other plant responses are the products of natural selection. So, of course, are various animal re-

A Question of Sensitivity

sponses. Bodily shapes in both plant and animal kingdoms are also products of natural selection.

Evolution has produced some striking similarities between animal and plant living strategies. However, the basic direction of evolution in the two kingdoms has been different.

Animals depend on plants or other animals for food. Most animals move from place to place to get food. Since many plants can regenerate themselves if just small pieces of them remain, most animals are more easily killed than plants.

Everything seems to happen more quickly for animals than for plants. So animals have developed swifter, more involved information systems than plants.

Animals have acquired specialized organs for sensing their environment. They have developed specialized organs for communicating and interpreting sensed data. In other words, animals have developed nervous systems.

The complex nervous systems of higher animals include mechanisms for forming visual images and for distinguishing various sounds. They also include mechanisms for identifying scents, for detecting temperature changes, and for acting swiftly on the information collected.

Higher animals have also developed emotional reactions such as fear and sexual drives. These emotional reactions motivate the animals to act on the information they receive.

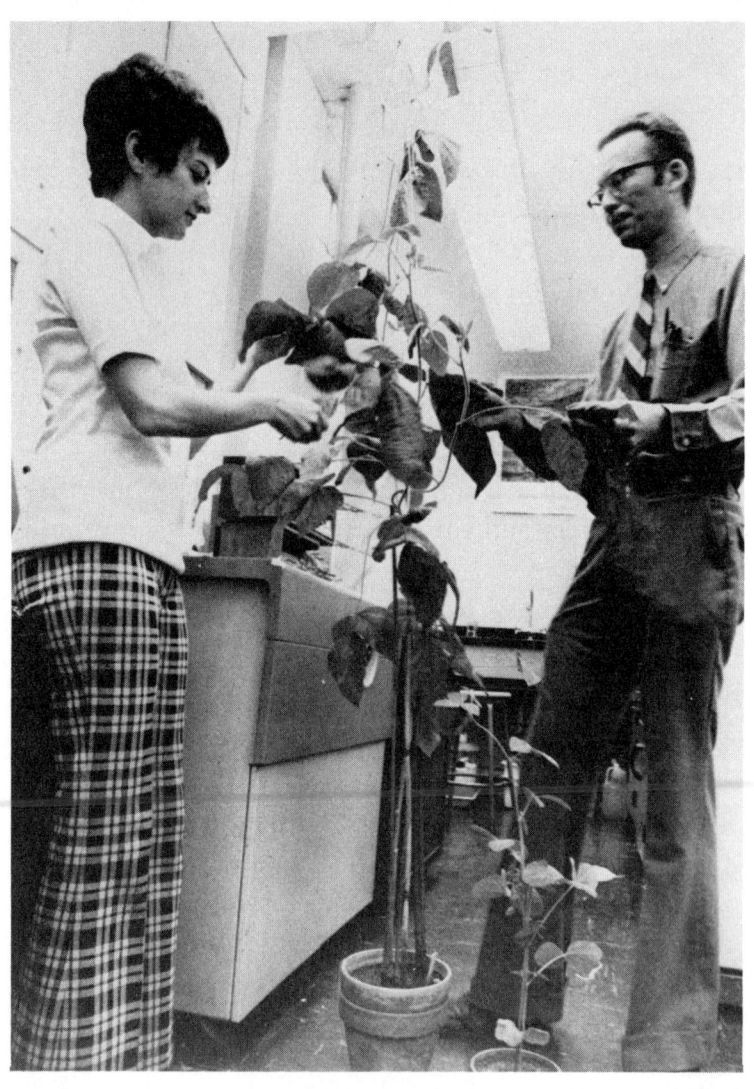

Botanist Larry Nooden (right) of the University of Michigan studied plant responses to various stimuli. In 1974, he concluded that plants, like people, occasionally commit suicide. Dr. Nooden also stated that the plants send out a "death signal" before they die. (*UPI*)

A Question of Sensitivity

Many, perhaps most, nonhuman animal reactions appear to be instinctual. That is, those reactions are mostly automatic—they are not the result of conscious, deliberate thought.

However, animal reactions do seem to involve a certain kind of awareness. Occasionally, nonhuman animals even seem to deliberate about a course of action. They certainly seem to have emotions. Some higher animals are able to learn tasks. They are also able to learn to communicate in limited, nonverbal ways with humans.

It is from this animal awareness, scientists point out, that human consciousness and human emotions have evolved. Our consciousness and our emotions have their biological base in the animal bodies we have inherited from our nonhuman ancestors.

In plant bodies, there appears to be no biological basis for even the limited awareness that some invertebrates (animals without backbones) seem to have. No kind of nervous system has been found in a plant. Plant responses are not in any sense voluntary.

Plants have no way of recognizing individuals, either other plants or animals. Plants do not appear to feel pain. They have no central organ in any way like a brain to receive pain impulses. There is no evidence that growth responses signal any sense of well-being in plants. Plants' responses appear to be entirely unconscious.

You may wonder why scientists say "appear to be." Can't they make up their minds?

PLANT MYSTERIES

Scientists do not expect to be absolutely certain of anything. They are not absolutely certain that a plant feels nothing at all.

We have already said that there are some apparent resemblances between the mechanism that controls the sleep movements of plant leaves and the mechanisms of animal nerve transmissions. Both types of mechanisms involve the movement of ions (charged atoms) through tissue pores. Both depend on transfers of potassium ions.

Such a mechanism also seems to control the sudden response of the sensitive plant, *Mimosa pudica*, to touch. The sensitive plant gives off an electrical signal when it suddenly closes its leaves. The leaves of the sensitive plant also fold up at night like those of numerous other plant species.

What's more, the sensitive plant can be put to sleep!

How? By chloroform! When a sensitive plant is enclosed in a jar with a small quantity of chloroform, it soon becomes insensitive. It will not respond to touch. But when the chloroform wears off, the sensitive plant will once again fold its leaves when touched.

Because of this, scientists can't be absolutely certain that a plant feels nothing at all. But scientists can be sure that, by comparison with most animals, plants don't feel very much.

Plants don't understand words. They can't sense the love or good feelings of the person who takes care of them. They don't pay attention to animals or other

A Question of Sensitivity

plants. They do not have the biological equipment to do these things.

Plant bodies are not specialized for such responses. Plant bodies have evolved to respond to light, gravity, contact, and certain chemical and heat gradient levels. Plant bodies seem to respond to electrical fields. They may respond to sound. However, plants do not have even the limited ability of some nonhuman animals to respond to symbols, thoughts, or emotions.

It is particularly unreasonable to credit plants with occult powers, such as the ability "to read minds" and telepathic communication. There is no scientific evidence that humans, with their highly developed consciousness, can do such things. So why should we expect it of plants? If there is such a thing as plant consciousness, it must be more primitive than that of the simplest animal with a central nervous system. *Certainly there is no evidence to show that plants are conscious in any sense of the word.*

Why then do so many people think that talking to their plants makes the plants grow better? Can talking to plants possibly do any good?

Several prize-winning gardeners have been known to chatter away to their blue-ribbon fruits and vegetables. But, many other expert gardeners and horticulturists never say a word to their plants. And those plants also flourish and win prizes.

Plants do not need conversation, but they do need

PLANT MYSTERIES

care and attention. So, in one sense, talking to plants may help. It may make a gardener more aware of the growing conditions and general health of his or her plants.

Plants need attention. They require the particular light, water, and temperature and moisture conditions under which they grow best. These requirements vary from plant species to plant species. They also vary from location to location.

Houseplants in a window should be turned frequently. Some plants should be pinched back periodically. Plants should be repotted when they threaten to outgrow their containers. House and garden plants must be watched for signs of disease and pest infestation. They also need to be fertilized.

While plants don't need affection, they may need nurturing. Talking to plants may make them grow better. But that's not because the plants understand the conversation or absorb waves of love and tenderness from the person talking to them. It's because the act of talking focuses the attention of the gardener on the plants. Many people need to talk to plants in order to be fully aware of the needs and growing conditions of those plants.

Suggested Further Readings

Baker, Jerry. *Talk to Your Plants*. New York: Pocket Books, 1974.
Budlong, Ware T. *Performing Plants*. New York: Simon and Schuster, 1969.
Coulter, Merle C., and Dittmer, Howard J. *The Story of the Plant Kingdom*. Chicago: University of Chicago Press, 1964.
Darwin, Charles. *The Power of Movement in Plants*. (Various reprints available).
Fogg, G. E. *The Growth of Plants*. Baltimore: Penguin Books, 1963.
Galston, Arthur W. *The Life of the Green Plant*. Englewood Cliffs, N.J.: Prentice-Hall, 1964.
Hutchins, Ross E. *Strange Plants and Their Ways*. New York: Rand McNally, 1958.
Klein, Richard M., and Klein, Deanna T. *Discovering Plants*. Garden City, N.Y.: Natural History Press, 1968.
Kramer, Jack. *Plants That Play Games*. New York: Collins and World, 1977.
Milne, Lorus, and Milne, Margery. *The Nature of Plants*. Philadelphia: J. P. Lippincott, 1971.
Northern, Henry, and Northern, Rebecca. *Ingenious Kingdom*. Englewood Cliffs, N.J.: Prentice-Hall, 1970.

PLANT MYSTERIES

Overbeek, Johannes, and Wong, Harry K. *The Lore of Living Plants*. New York: Scholastic Book Services, 1964.

Poling, James. *Leaves, Their Amazing Lives and Strange Behavior*. New York: Holt, Rinehart and Winston, 1971.

Ray, P. M. *The Living Plant*. New York: Holt, Rinehart and Winston, 1972.

Selsam, Millicent E. *Plants That Move*. New York: William Morrow and Co., 1962.

Tompkins, Peter, and Bird, Christopher. *The Secret Life of Plants*. New York: Harper and Row, 1973.

Went, Frits W. *The Plants*. New York: Time, Inc., 1963.

Index

Abscissic acid, 126, 127
Acetylecholine, 120
African violet, 65, 67, 68
Agar, 102
Agro-Lite fluorescent lamp, 107
Allard, H. H., 113
American Association for the Advancement of Science, 80, 81
Anchor College of Truth, 58
Animal responses, 98
Annamalai University, India, 129
Aphids, as plant pests, 19
Apical buds, 121
Apple trees, 92
Argosy magazine, 47
Asexual reproduction, 96
Ashley, Terry, 94, 95
Auxin, plant hormone, 102–106, 109, 110, 121, 122

Backster, Cleve, 34–59, 65–75, 79, 80, 89, 133
"Backster Effect," 47

Baker, Jerry, 22, 149
Baltimore Sun, The, 27, 51
Bean plant, 125; Climbing behavior, 14
Bedford College, London, England, 103
"Be Kind To Your Plants" by Richard Martin, 47
Bird, Christopher, 48, 56–61, 65, 129, 131, 150
Bladderwort, carnivorous plant, 16
Bonner, James, 114
Borthwick, H. A., 115
Bose, Sir Jagadis Chandra, 59
Boysen-Jensen, P., 96, 103
Brassins, 105
Brine shrimp killing experiment, 42–46, 49, 52, 69, 76–78
Broman, Francis E., 136
Bunning, Erwin, 117
Burbank, Luther, 24, 25, 59

Camptosorus rhizophyllus, walking fern, 13, 14

Index

Carbon dioxide, 31, 92
Carnivorous plants, 14–18, 92, 94
Carotene, 106
Carotenoids, 106, 108
Carver, George Washington, 23, 24, 59
Cell growth, plant, 86, 98, 104, 112, 120, 121
Central Intelligence Agency (CIA), 37, 38
Chemistry, journal, 71
Chemotropic response, 111
Clairvoyance, 46
Cocklebur, 114, 115
Coconut milk, 123
Coleoptile, 94, 101
Coleus, 85, 136
Collier, James Lincoln, 51
Cornell University, 76
Costa Rica, 85
Cytokinins, 123, 125

Darwin, Charles, 12, 24, 59, 88, 93, 100, 117
Darwin, Francis, 93, 100
Day-neutral plant flowering category, 92
Dean, Douglas, 51
Dial-A-Plant, New York Telephone Co., 26
Digestion, plant, 98
Dionaea muscipula, Venus's flytrap, 14, 17, 72, 89
"Do Plants Have Feelings, Too?" magazine article, 27, 28
Dormancy of seeds, 91

Dracaena massangeana, a houseplant species, 38, 40

Earth's gravitational field, 88
Electrical activity in plants, 32–34, 104–109, 118, 120
Electrical resistance of skin, 35, 37
"Electrical Responses of Plants to External Stimuli," 80, 81
Enzymes, 98, 126
Esser, Aristide, 51
Extrasensory perception (ESP), 46, 52, 58, 74

Faraday cage, 52, 76
Ferns, sprouting roots from leaf tips, 13, 14
Fort Holabird, Maryland, 37

Galston, Arthur W., 79, 81, 104, 116, 117, 124, 129
Galvanic skin response, 37
Galvanometer, 50, 51
Garner, W. W., 113
Gasteiger, Edgar L., 76, 77, 80
Genes, 88
Gennaro, Joseph F., Jr., 94, 95
Geotropism, negative—growth of plant stems against gravity, 87
Geotropism, positive—growth of roots in response to gravity, 87, 104, 108, 109
Germination of seeds, 89, 91, 124, 126, 134
Gibberellin, 124–126
Grahm, Lennart, 105

154

Index

Grandma Putt, 22, 23
"Greenthumb people," 21

Hageseth, Gaylord T., 133, 134
Hall, Manly P., 25
Hammer, Karl C., 114
Harper's Magazine, 56, 65, 70, 72
Hendricks, S. B., 115
Hertz, E. H., 105
Hormones, 98–102, 110–112, 120–125, 127
Houseplants as pets, 16, 26
Human emotions, 145
Huxley, Sir Julian, 129
Hydrilla, 129
Hydroids, sea animals, 108
Hydrotropism, root growth toward water, 87
Hypnotism, 58

International Journal of Parapsychology, The, 46, 72
Ions, 32, 120

Jaffe, M. J., 119

Kinetin, 124
Kinins, natural, 124
Kmetz, John M., 77–80

Lawrence, L. George, 55–58, 60
Legumes, 23
Lie detector, polygraph, 34–40, 46–52, 65, 76–78
"Life and Light" by George Wald, 106

Long-day, plant flowering category, 92, 114, 115
Lord, Arthur, 136
"Love Among the Cabbages" by Tompkins & Bird, 56, 57, 65
"Lunar Day," period determined by motion of moon, 111
Lund, E. J., 104, 105

McGraw, Walter, 47
Marine animals, 84, 129
Martin, Richard, 47
Meisner, Gerald W., 134
Membranes, cell, 106
Metabolism, 98
Milstein, George, 133
Mimosa pudica, the sensitive plant, 9–11, 82, 89, 146
Mitchell, John W., 125
Monochromatic light, 106
Moon flowers, 89
"More Experiments in Electroculture" by L. G. Lawrence, 55
Morning glories, 89
Morrison, Irma L., 134
Morrison, Ralph M., 134
Music, to stimulate plant growth, 27, 129–133, 136, 137
"Music to Grow Plants By," record, 133, 135

Nastic movements, 89
National Wildlife magazine, 47
Natural History magazine, 117
Negative phototropism, 85

Index

Nervous response, animal, 111, 112, 143
Nitrogen, essential plant nutrient, 23, 92
Nooden, Larry, 144
North Carolina, home of Venus's flytrap, 14
Nutation, plant movement, 87, 88

Ohio University, 119
One-celled animals, experiments with, 52

"Paranormal matrix," 58
Parapsychology, 57
Peanut plant, 23, 24
Philodendron, 51, 56
Philosophical Research Society, 25
Photoperiodism, 92
Phototropism, plant growth toward light source, 85, 87, 99, 104, 106, 109
Photosynthesis, 92
Phytochrome, 115, 117, 118
Pickard, Barbara, 33
Pitcher Plant, *Saracenia purpurea*, 18
Plant; diseases, 32; electrical activity, 32, 33, 71; killing experiment, 49, 50; respiration, 31, 32; sap circulation, 31, 32
"Plants Are Only Human" by Walter McGraw, 47
Poinsettia, 65, 68
Pollen grain, 110

Polygraph, lie detector, 34–40, 46–52, 65, 76–78
Popular Electronics, magazine, 55
Potassium, 119
"Primary perception" in plants, 41, 42, 46, 52
Protoplasm, 86, 129, 130
Putt, Grandma, 22, 23

Racusen, Richard, 119
Radioactive fruit flies, 95
Radioactive phosphorus, 95
Raga, traditional Indian music, 129
Ray, Thomas S., Jr., 85
Reader's Digest, 28
Regeneration, plant, 143
Repulsive odors, plant defenses, 142
Respiration, plant, 31, 95, 96
Retallack, Dorothy, 136–140
"Rhapsody In Blue" by George Gershwin, 131
Rhododendron, 89
Rockland State Hospital, New York, 51

Sap circulation in plant tissues, 31, 32
Sarracenia purpurea, pitcher plant, 18
Satter, Ruth, 119
Sauvin, Paul, 56, 57, 60
Schrank, A. R., 105
Schwebs, Ursula, 65–75
Scientific American, 106
Seaweed, 84

Index

Secret Life of Plants, The, 57, 59, 61–63, 129, 135, 136, 150
Seeds, traveling, 84
Sensitive plant, *Mimosa pudica*, 9–11, 82, 89, 146
"Sensitive Plant, The" poem, 11
Sexual reproduction, 95
Shelley, Percy Bysshe, 11, 12
Short-day, plant flowering category, 92, 113–115
Singh, T. C., 129–131
Skototropism, plant growth toward dark object, 85, 87, 109
Smith, George E., 131, 132, 136
Sodium, 119
Sound of Music and Plants by Dorothy Retallack, 139
South Carolina, home of Venus's flytrap, 14
Soviet Research, 57
Soybean plants, 131
Spider mites, as plant pests, 19
Sprouting, 91
Statoliths, 109
Steward, F. C., 122–124
Strong, Donald R., Jr., 85
Sundew, carnivorous plant, 95
Sunflowers, 88

Talk to Your Plants by Jerry Baker, 22
Tendrils, 109
Telepathy, 46
Theory of evolution, 12, 24
Thigmotropism, plant growth movement in response to touch, 109

Tobacco plant, 113
Tompkins, Peter, 48, 56–61, 65, 129, 131, 150
Tropisms, growth movements of plants, 84–89, 97, 111, 137
Turnip seeds, 134
Turgor movements, in plants, 89
Tuskegee, Institute, Alabama, 24

U.S. Department of Agriculture, 115, 125

Vacuole, 86
Veena, Indian stringed instrument, 130
Venus's flytrap, 14, 17, 72, 89
Vines, 87, 109
Vision, 108
Visual purple pigment, 108
Vitamin A, 106, 108
Vogel, Marcel, 57, 59, 60
Volt-Ohmmeter, 65–69

Wald, George, 106, 108
Wall Street Journal, The, 47
Washington-Lee High School, Arlington, Va., 70
Weinberger, Pearl, 132, 133, 139
Went, Frits, 103, 150
Westinghouse Laboratories, 108
Witch hazel, 84
"Wizard of Tuskegee," G. W. Carver, 23

X-rays, 52

ABOUT THE AUTHORS

Michael J. Cusack was born in Cork, Ireland. He attended Columbia University and Fairleigh Dickinson University. He is presently editor of *Science World*, Scholastic Magazines, Inc.

Mr. Cusack has written many articles on science, many of which have received writing awards.

Mr. Cusack and his wife, Anne, live on Roosevelt Island, New York City, with their three daughters, Deborah, Deirdre, and Jennifer.

Anne E. Cusack was born in Hacksensack, New Jersey. Graduating from Seton Hall University with a B.S. in English, she was awarded a Woodrow Wilson Fellowship at Stanford University.

For three years, Ms. Cusack taught English and creative writing at a girls' private high school. Later, she devoted most of her attention to raising a family and writing. Her poems have been published in numerous "small" magazines. Several of her articles and a science-fiction short story have also been published. Ms. Cusack wrote two filmstrip scripts and several teaching guides for the award-winning audio-visual series, *Human Issues In Science*, published by Scholastic Magazines, Inc., in 1975.